LIFE UNDER A CLOUD

LIFE UNDER A CLOUD

American Anxiety About the Atom

ALLAN M. WINKLER

New York   Oxford
OXFORD UNIVERSITY PRESS
1993

Oxford University Press

Oxford New York Toronto
Delhi Bombay Calcutta Madras Karachi
Kuala Lumpur Singapore Hong Kong Tokyo
Nairobi Dar es Salaam Cape Town
Melbourne Auckland Madrid

and associated companies in
Berlin Ibadan

Copyright © 1993 by Allan M. Winkler

Published by Oxford University Press, Inc.,
200 Madison Avenue, New York, New York 10016

Oxford is a registered trademark of Oxford University Press

Library of Congress Cataloging-in-Publication Data
Winkler, Allan M., 1945–
Life under a cloud: American anxiety about the atom
Allan M. Winkler.
p. cm. Includes bibliographical references (p.   ) and index.
ISBN 0-19-507821-7
1. United States—Military policy.
2. Nuclear weapons—United States—History.
3. Nuclear engineering—Government policy—United States—History.
I. Title. UA23.W485 1993 355'.0335'73—dc20 92-20013

I gratefully acknowledge those who granted permission to quote from the following material.

From *The Poetry of Robert Frost*, edited by Edward Connery Lathem. Copyright 1947, © 1969 by Holt, Rinehart and Winston. Copyright © 1975 by Lesley Frost Ballantine. Reprinted by permission of Henry Holt and Company, Inc.

From the songs "The Wild West Is Where I Want To Be," copyright © 1953 by Tom Lehrer, copyright renewed; "We Will All Go Together When We Go," copyright © 1958 by Tom Lehrer, copyright renewed; "Who's Next?," copyright © 1965 by Tom Lehrer; "So Long, Mom," copyright © 1965 by Tom Lehrer. All used by the permission of Tom Lehrer.

From "Fall 1961" from *Selected Poems* by Robert Lowell. Copyright © 1969, 1970, 1973, 1976 by Robert Lowell. Reprinted by permission of Farrar, Straus & Giroux, Inc.

From *Li'l Abner*, written and produced by Norman Panama and Melvin G. Frank, in New York City, on Broadway at the St. James Theater, October 1956. Used by permission of Norman Panama.

From "What Have They Done to the Rain." Words and music by Malvina Reynolds. Copyright © 1962 by Schroder Music Co. (ASCAP), renewed 1990. Used by permission. All rights reserved.

From the album *Atomic Cafe*. Used by permission of Rounder Records.

From "Dig a Hole in the Ground, Or, How to Prosper During the Coming Nuclear War." Copyright © 1982 by Pine Barrens Music (BMI). Reprinted with permission from *Breaking from the Line: The Songs of Fred Small* (Yellow Moon Press, 1986).

From "Two Suns in the Sunset," by Roger Waters. Copyright © 1982 by Pink Floyd Music Publishers, Inc.

2 4 6 8 9 7 5 3 1

Printed in the United States of America
on acid-free paper

For Sara

Acknowledgments

In working on this book, I have enjoyed all kinds of assistance. A full-year fellowship from the National Endowment for the Humanities in 1981–1982 allowed me to travel to archival collections and to devote full time to research. Summer grants from the University of Oregon in 1980 and from the Oregon Committee for the Humanities in 1984 were similarly useful in providing time to read the vast literature about nuclear affairs. When I was ready to write, Miami University of Ohio gave me a much-appreciated semester off in 1990 and so helped me get started. Other kinds of support were equally valuable. I appreciate the aid of Michael O'Brien, a colleague and friend, who shared his computer expertise as I began to organize my material. I am also grateful for the encourgement of friends in the nuclear-history field—J. Samuel Walker, Martin J. Sherwin, and Gregg Herken among others—and for the support and close reading of the manuscript I received from another friend, Paul Boyer. Over the years, my sister, Karen Winkler Moulton, has been my most perceptive critic, and I would like to give her a special word of thanks both for helping me think through the argument and for editing the prose. As always, I appreciate the support of Gerry McCauley, who promoted this project and found it a home. Thanks, too, to Jeri Schaner, my secretary, who assisted me in countless ways, and to Scott Lenz, my editor at Oxford University Press, who prepared the manuscript for publication. My children, Jenny and David, have been a constant source of pleasure as their lives have intertwined with mine and kept my work in perspective. Finally, I am grateful to Sara Penhale, to whom this book is dedicated, for the support she has given over the past several years, first as my companion and now as my wife.

Oxford, Ohio   A. M. W.
August 1992

Contents

Prologue

An aging tortoise, eager for a cozy home, crawled into a woodpile being readied for an autumn bonfire. The tortoise must have felt increasingly secure as each new stick provided further protection from wind and rain. Then one day, the pile-builders lit the bonfire, incinerating both the wood and the family pet.

Nicholas Humphrey, an English experimental psychologist, sadly recalled the fate of his tortoise Ajax as he looked back at the history of the atomic age in a commemorative lecture in 1981. Interested professionally in the evolution of social consciousness, Humphrey had become increasingly concerned with questions of war and peace and their human implications. Now he referred back to Ajax as he asked, "Are there not some of *us* who still believe that the piling up of weapon upon weapon *adds* to our security— that the dangers are nothing, compared to the assurance they provide?" Humphrey was speaking over the BBC in an address honoring Jacob Bronowski, the eminent anthropologist and creator of the popular television series *The Ascent of Man*, who had visited Nagasaki in the fall of 1945 and found "civilization face to face with its own consequences." Humphrey, assessing the world's subsequent development, was unable "to avoid the conclusion that mankind, having flown too near the sun, is already in a stall." Humphrey's voice was but one of many that could be heard in the midst of a worldwide wave of anti-nuclear agitation. In Great Britain, France, Germany, and particularly the United States, activists, ranging from baby-toting housewives to internationally known physicians, protested the buildup of atomic arsenals that threatened the very survival of the human race. But this was not the first wave of anti-nuclear protest. American critics of atomic policy, more than any of the world's citizens, had been in the forefront of the protest movement in launching a series of earlier attacks on policies they felt threatened world stability. They had challenged the tests that made "fallout" a household word in the 1950s. They had questioned the defense policy that resulted in near-cataclysmic confrontation in the Cuban missile crisis in the 1960s. And now in the 1980s they were speaking out once again.[1]

In each case, the challenge had only a passing impact. In the 1950s, protesters persuaded the American government to accept a voluntary testing moratorium. In the 1960s, pressure promoted the Limited Test Ban Treaty of 1963 that banned atmospheric explosions. In the 1980s, anti-nuclear arguments led President Ronald Reagan to soften his Cold War rhetoric and finally to accept a treaty banning intermediate-range nuclear weapons.

But each time, after marginal success, anti-nuclear activism disappeared. Although serious problems still demanded solutions, most Americans—and their counterparts in countries possessing nuclear weapons—seemed unwilling to confront the issue further. Anxieties remained but could not mobilize a public response. *Life Under a Cloud* seeks to explain why deep-rooted and corrosive fears of nuclear destruction have failed in the past fifty years to bring atomic weaponry under effective control.

Fear, tempered by hope for a brave new atomic future, is part of our past and present cultural landscape. Scientists working on the Manhattan Project to create an atomic bomb during World War II were afraid of what the new weapon might do. Similarly, President Franklin D. Roosevelt perceived the catastrophic possibilities ahead. Secretary of War Henry L. Stimson, overall director of the project, became increasingly aware of the bomb's potential impact on international diplomatic affairs. In 1947 he reflected that "with the release of atomic energy, man's ability to destroy himself is very nearly complete." As hydrogen weapons replaced atomic bombs in the 1950s and kilotons gave way to megatons in the decades that followed, such fear grew even more pronounced, and speculation about a dismal or nonexistent future became more common. Scientists in the 1970s and 1980s predicted deadly epidemics of radiation-related illnesses, devastating climatic adjustments, and the death of life as we know it. Biologist Lewis Thomas observed in 1981 that "the fossil record abounds with sad tales of creatures that must have seemed stunning successes in their heyday, wiped out in one catastrophe after another. The trilobites are everywhere, elegant fossil shells, but nowhere alive. The dinosaurs came, conquered, and then all at once went." We, too, now faced extinction, he suggested, for which we alone would be responsible: "We will not be wiped off the face of the earth by hard times, no matter how hard. We are tough and resilient animals. Good at hard times. If we are to be done in, we will do it ourselves by warfare with thermonuclear weapons."[2]

Fear, however, was muted by hope. With the advent of nuclear energy, a new age beckoned, and for years the possibilities appeared boundless. Atomic energy could "usher in a new day of peace and plenty," according to University of Chicago chancellor Robert M. Hutchins in 1945. This wonderful new force, Walt Disney proclaimed a decade later in the popular children's book *Our Friend the Atom*, could "be put to use for creation, for the welfare of mankind." Despite near-catastrophic accidents around the world, the administration of George Bush began at the end of the 1980s to consider the advantages of nuclear power again.[3]

Even so, fear remains an undeniable part of our nuclear age. Yet throughout the atomic era, many Americans have refused to face up to fear in a lasting way. Unless issues became immediate— with children threatened by residual fallout or the future compromised by the imminent prospect of war—most citizens of the United States have avoided thinking about vexing nuclear controversies in the naive hope that problems will simply disappear. In 1954, 63 percent of a national sample in a poll taken by George Gallup and the American Institute of Public Opinion worried that the hydrogen bomb would be used against the United States in a world war. But in a another Gallup poll taken the same week, 95 percent of the respondents said they had not considered moving elsewhere to escape attack. They found it simpler to cast furtive, sidelong glances at the bomb, just as we do today, and so avoid facing it directly.[4]

Fear needs to be more sharply focused before it produces a response. A cultural overview of the atomic age indicates that individuals do affect the course of history. In the past half-century, outspoken individuals have highlighted the nuclear issue and so encouraged constructive protest. Literary figures, in particular, have had a significant impact on shaping public response to atomic energy. John Hersey, in *Hiroshima*, first published in *The New Yorker* in 1946 and then in book form, horrified readers with his dispassionate but still vivid portrayal of damage. In an age before television became a common fixture, he provided the images that showed the force of the bomb and made readers ask about future uses of this potent new weapon. Nevil Shute proved equally frightening with *On the Beach*, a novel published in 1957 and then made into a film which described a war that wiped out life on Earth. Shute's work helped focus attention on fallout and encouraged the editorials, public-interest advertisements, and rallies that finally led to limits on nuclear testing. Jonathon Schell had an even more dramatic impact with *The Fate of the Earth*, which, like *Hiroshima*, was published first in *The New Yorker* and then in book form in 1982. Its riveting images of nuclear destruction prompted sermons, discussions, and debates throughout the United States; inspired demonstrators; and helped spark support for a nuclear freeze and other accords.

Protests, whether orchestrated by such incisive individuals as Jonathon Schell or launched in response to unfortunate events like the Cuban missile crisis, have at times affected policymakers. Dwight D. Eisenhower reluctantly bowed to public fears of fallout and announced a voluntary testing moratorium in 1958. Recognizing the public concern about possible nuclear holocaust resulting from his provocative posture toward the Soviet Union over Cuba in 1962, John F. Kennedy responded by supporting a Limited Test Ban Treaty to alleviate anxieties. Ronald Reagan, undoubtedly troubled about the national and international response he provoked by his program of massive spending for defense and his hard-line nuclear negotiating stance, shifted course in his second term and so provided the opening that resulted in rapid acceptance of the treaty banning intermediate-range nuclear missiles.

In each case, the response of policymaking officials has quelled public fears and quieted the national debate. Eisenhower's voluntary testing moratorium silenced criticism for a time; Kennedy's test ban treaty effectively checked fears of nuclear war for the next fifteen years; and Reagan's about-face undermined a wave of anti-nuclear protest in the mid-1980s.

The hopes and fears, the challenges and diplomatic initiatives, have all unfolded within the framework of an ongoing dialogue among scientists, cultural critics, and government officials. This book examines these groups and their shifting influence in the United States. Scientists spoke out first as the experts who had unleashed the awesome new force and best understood how to deal with it. Though many had doubts from the start about the monster they might create, they set aside their anxieties in the interests of defeating Adolf Hitler and his Axis allies. "How Well We Meant"—so Nobel Prize winner I. I. Rabi titled his speech to Los Alamos colleagues at a reunion several decades after their first success. Meanwhile, literary and artistic commentators began to explore the dramatic possibilities of atomic holocaust. "When the bomb was dropped," author Isaac Asimov noted, "atomic-doom science-fiction stories grew to be so numerous that editors began refusing them on sight." Novels, stories, comics, films, songs— all served as something of a safety valve, allowing fears to find expression as artists indulged their creative vision. But government officials rather than scientists or cultural critics seized the initi- ative in shaping the public agenda. The scientists and intellectuals, policymakers felt, failed to understand political demands. As stra- tegic and military planning came to dominate national discourse, they argued that they alone had the expertise to protect the nation from nuclear threat. Their reach extended in all directions. They helped promote public dreams about nuclear planes and ships, about the medical benefits of atomic isotopes, and about the ex- travagant possibilities of nuclear power. They made the final de- cisions about how development should proceed.[5]

   Life Under a Cloud argues that the various groups operated in a series of intersecting circles. Members knew one another, spoke to one another, and sought to persuade one another as they inter- acted over time. Cultural commentators watched the scientists carefully as they devised ever more sophisticated weapons. But cultural criticism has only occasionally had an impact on American political life. Elected and appointed officials, who encouraged sci- entists to pursue nuclear research and funded their work, were cognizant of both cultural and scientific critiques, even when they found such complaints frivolous. Listening to criticism, they shifted course only when pressure became intense and, more often, simply devised ways to parry nuclear fears. Frequently they used public relations to manipulate national response, as when Dwight Eisenhower's administration launched a campaign to discredit the end-of-the-world scenario in the book and film versions of *On the Beach*. Or they used limited agreements to launch new weap- ons research, as when John Kennedy allowed increased testing as long as it occurred underground and did not pollute the atmosphere. Successful in their efforts, they defused most challenges to their approach. Single-minded about promoting strategic security, whatever the cost, they maintained their own commitment to proceed with nuclear development and thus avoided a searching reexamination of American policy.

I have chosen to tell this story in a series of topical chapters moving chronologically through the past fifty years. In highlighting the key moments in the American struggle to come to terms with atomic energy, I have focused on the ongoing dialogue among the various participants over such issues as military strategy, civil defense, and nuclear power. Often debate over one issue took place at the same time as discussion about another, and I have tried to show these connections when they occurred, but have still sought to explore each problem as fully as possible in a chapter of its own. Though I have occasionally looked back to record the origins of conflict over a particular question, for the most part I have tried to move the narrative steadily forward in considering the major national issues as they emerged.

This book, then, describes the origins of the framework within which we consider nuclear issues today. It is a study of reactions, an examination of how opinion developed and intertwined with policy, an intellectual and cultural survey of how attitudes and ideas affected and were affected by political decisions over the past fifty years. It asks what caused nuclear fears to focus and what caused them to abate. Though recent events in eastern Europe have eased global tensions, the threat of atomic annihilation remains bound up with the patterns of our nuclear past. In detailing the delicate balance between hopes and fears, this book shows how Americans have wrestled, and still wrestle, despite the end of the Cold War, with nuclear issues in our continuing national debate.

1

Origins of the
Atomic Age

The atomic bomb revolutionized American life. In all areas—
economic, social, political—it challenged old assumptions and
forced reconsideration of accepted standards. The process of development established an entirely new model for collaboration between the worlds of science and government. The Manhattan
Project mobilized scientists in the largest venture attempted up to
that time and drew them permanently into the realm of public
affairs. Scientists who had worked independently in the past now
found themselves involved in cooperative, and occasionally contentious, relationships with federal agencies and entangled in policy debates they had been able to avoid before. At the same time,
the bombs dropped on Hiroshima and Nagasaki raised new questions for government officials about military strategy and future
use of atomic weapons. Surpassing the most hopeful expectations
about potency and yield, the new bombs helped end the war and
so provided the framework for the fragile postwar peace, even as
they initiated an almost uncontrollable arms race. The bombs also
embroiled ordinary citizens, once they were informed, in long-
lasting discussions about the role of atomic energy in the postwar
world. Having released the long-trapped nuclear genie, Americans

had to face new issues of life and death and human morality that could not be ignored. What role did the bomb have in diplomatic affairs? Who should make the political decisions about the subsequent development of atomic energy? And how might the benefits of this extraordinary new force offset its terrible destructive possibilities? Those questions, still pertinent today, first surfaced in the 1940s, and the answers reached then helped shape the patterns of the next fifty years.

Scientists had long speculated about hidden sources of energy that might somehow be released. The discovery of radioactivity in 1896 opened the way to exploration and experiment that culminated half a century later in the crash program to create an atomic bomb. The light emitted by thorium and uranium attracted little interest until Polish-born chemist Marie Curie, working on her doctoral dissertation in France, isolated a new element, radium, which gave off even stronger rays. When, in 1900, she and her husband, Pierre, displayed active compounds that glowed with an eerie light, other scientists took note. Among those interested were two Englishmen, physicist Ernest Rutherford and scientific publicist Frederick Soddy. Their experiments persuaded them that radioactivity reflected basic changes within matter. If the energy within matter could be tapped, the enthusiastic Soddy wrote, "the future would bear as little relation to the past as the life of a dragonfly does to that of its aquatic prototype." Those who learned to transmute matter "could transform a desert continent, thaw the frozen poles, and make the whole world one smiling Garden of Eden."[1]

Not all observers viewed the prospects so optimistically. Atomic energy might well be used instead to create weapons of war. In *The World Set Free*, a novel published in 1914 and dedicated to Soddy, English author H. G. Wells explored the notion of a radioactive chain reaction that could be used in "atomic bombs." Looking ahead to the 1950s, he imagined a volatile element called Carolinum and a continuing explosion that, once launched, could last for weeks or months or years. With social and political sensibility less sophisticated than scientific expertise, the world Wells imagined was soon out of control. "Power after power about the armed globe," he wrote, "sought to anticipate attack by aggression. They went to war in a delirium of panic, in order to use their bombs first." Each bomb was potent enough to devastate its target.

Uncontrollable waves of fire released "puffs of luminous, radio-active vapour drifting sometimes scores of miles from the bomb centre and killing and scorching all they overtook." With humanity close to extinction as war ran its course, weary survivors finally came to their senses and seized upon the promise of the atom to create a new society in which the blessings of technology could be enjoyed. Wells, with his vivid imagination, took the discoveries of early twentieth-century science a step further than the physicists and so initiated a speculative dialogue that became even more intense as the nuclear age unfolded.[2]

In the 1930s, scientists began to consider atomic possibilities more seriously. Leo Szilard, born in Hungary, educated in Germany, and living in Great Britain, was one of the first to reflect on how the atom might be split. Energetic, imaginative, intent on saving the world, he had studied physics with Nobel laureate Albert Einstein and other theoreticians in Berlin. Inventive, he had applied for twenty-nine German patents, either individually or jointly with Einstein, starting in 1924, until Germany became inhospitable to those of Jewish ancestry and he fled a day before departure became more difficult. "You don't have to be cleverer than other people," he remarked, "you just have to be one day earlier than most." In England, he became intrigued with work done by Rutherford and by Danish physicist Niels Bohr on the nucleus of the atom. He began to believe that a chain reaction might be possible if a neutron—a particle with no electric charge—from one atom should strike another atom and release still other neutrons to continue the process. A report by Rutherford in the *Times* late in 1933 caught his attention, despite its disclaimer that "anyone who looked for a source of power in the transformation of the atoms was talking moonshine." Once engaged with the idea of a chain reaction, Szilard applied for a patent for that process, which he assigned to the British Admiralty in order to keep it secret.[3]

Other scientists proceeded systematically to explore the notion of nuclear fission. In Germany, in 1938, as similar experiments were being done in France and the United States, Otto Hahn and Fritz Strassmann found radioactive elements present when they bombarded uranium with neutrons. Lise Meitner, an Austrian colleague, confirmed that fission—an entirely new reaction splitting the nucleus—had occurred. Informed of these striking de-

velopments, the eminent Bohr shared the news with other physicists as he spent several months at the Institute for Advanced
Study in Princeton. In the year following the first publication
announcing fission, nearly a hundred articles appeared in journals
around the world.[4]

Enrico Fermi, an Italian Nobel laureate working at Columbia
University, was one of those intrigued. Passionate about physics
since the age of fourteen, impatient with a purely theoretical approach, Fermi thought in concrete terms and was forever eager to
quantify what he saw. "Fermi's thumb," his wife wrote later, "was
his always ready yardstick. By placing it near his left eye and closing
his right, he would measure the distance of a range of mountains,
the height of a tree, even the speed at which a bird was flying."
He loved to classify, she added, "and I have heard him 'arrange'
people according to their height, looks, wealth, or even sex appeal."
Like other scientists who found Europe inhospitable, Fermi had
fled Italy when anti-Semitic decrees threatened his Jewish wife
and family. Once, while working in the Columbia physics building,
he cupped his hands together to form a ball as he considered the
possibilities of fission. A bomb like that, he said, could cause
Manhattan to disappear. Even so, he felt that there was only a
remote chance, "perhaps ten percent," that a controlled chain
reaction would work.[5]

Fermi and Szilard, who had also come to the United States,
began experiments in New York to test whether such a reaction
might be possible. They were encouraged by positive results yet
chastened by the military implications of their work as the world
hurtled toward war. Szilard in particular wished to alert the American government to the ominous possibility of an atomic bomb.

Albert Einstein was the logical choice to lend credibility to
such a warning. As early as 1907, he had suggested the equivalency
of matter and energy and had defined the formula $E = mc^2$ (energy
is equal to mass multiplied by the speed of light squared). After
outlining the general theory of relativity in 1915 and having it
verified in a solar eclipse in 1919, he was, a German magazine
noted, "a new figure in world history . . . whose investigations sig-
nify a complete revision of our concepts of nature, and are on a
par with the insights of a Copernicus, a Kepler, a Newton." Like
so many other European scientists who had faced anti-Semitic
persecution, the Jewish Einstein had fled Germany in 1932 and

had settled in the United States, where he had accepted an appointment at the Institute for Advanced Study.[6]

The warning, in the form of a letter, written by Szilard but signed by Einstein, was dated August 2, 1939, and addressed to Franklin Roosevelt. It mentioned experiments indicating that "the element uranium may be turned into a new and important source of energy in the immediate future." A nuclear chain reaction, releasing tremendous amounts of power, was possible, "and it is conceivable—though much less certain—that extremely powerful bombs of a new type may thus be constructed." The letter noted that German scientists were working in this area and suggested that the government assist American experimental efforts.[7]

The message made a difference. Alexander Sachs, an economist interested in science and with access to the White House, agreed to transmit the letter to the President. He saw Roosevelt in October, after the Nazi invasion of Poland brought the start of World War II. "This requires action," FDR told Edwin M. (Pa) Watson, his military aide, and he established a Uranium Committee, chaired by the director of the National Bureau of Standards. A year later, a National Defense Research Committee, headed by Vannevar Bush, president of the Carnegie Institution, assumed overall responsibility for fission experiments. Still, government support lagged until the summer of 1941, when the British government predicted that a bomb using pure uranium—or the recently discovered element plutonium—could cause a tremendous explosion and could be made within two years.[8]

With that prediction in mind, Roosevelt proposed a joint undertaking with Britain and established a new administrative structure. Bush, now director of the Office of Scientific Research and Development, remained deeply involved, but the War Department assumed direction in the fall of 1942. Brigadier General Leslie R. Groves was officer in charge; Secretary of War Henry Stimson had final responsibility for the effort. So started a process of collaboration between science, government, and the military establishment that dwarfed all prior efforts to develop new implements for war.[9]

The Manhattan Project was an enormous undertaking. Unfolding over a three-year period, the program cost more than two billion dollars, a massive sum of money at that time. It included thirty-seven different facilities in the United States and Canada and employed about a hundred twenty thousand people. Many of the nation's top nuclear physicists were involved. Ironically, *émigré* scientists, barred from other top secret research because of their foreign background, played a major role in this project, for government officials initially viewed it as far less important than other military development programs and, consequently, less of a security risk.[10]

Leslie Groves, coordinator of the effort, was a gruff but efficient career military officer. As a subordinate recalled, he was "the biggest sonovabitch I've ever met in my life, but also one of the most capable individuals." After graduating from West Point, where he had been known as "Greasy Groves," he had held desk jobs and directed construction projects (his latest, building the Pentagon) for much of his Army service. Anxious for an overseas command, the portly officer was disappointed with the Manhattan Project assignment. Only promotion from colonel to brigadier general made the mission more palatable.[11]

Groves found the project in disarray. If it was to succeed, he believed, scientists and government officials would have to be responsive to military needs. Quickly he moved to select sites for production plants, to acquire priority status for procurement of important materials, and to bring scientists into an organized administrative scheme. Concerned with security, he first sought to place scientists in uniform, then struggled to keep them compartmentalized so that each physicist knew only as much as was essential for his work. Scientists bristled at this first confrontation with government authority and insisted on their creative prerogatives. Recognizing finally that the project would fail without their total support, Groves backed down and permitted the scientists more flexibility, though he remained irritated at their lack of discipline and concerned that absolute secrecy be maintained.[12]

Constantly aware of their race against time, the scientists faced a daunting series of obstacles. They needed to verify that a chain reaction was indeed possible; they needed to isolate U-235, a rare uranium isotope, from the more plentiful U-238; and they needed to design the structure for a functional bomb.[13]

At the Metallurgical Laboratory at the University of Chicago, one group tried to produce a self-sustaining chain reaction. Led by Fermi, who had moved to Chicago, the scientists built a nuclear pile made of spheres of uranium oxide in a squash court under the west stand of Stagg Field, the university's football stadium. They began laying the spheres, encased in graphite bricks, on November 16, 1942. Two weeks later, on December 2, they were ready to withdraw cadmium control rods and observe whether the primitive reactor worked. Fermi, exhibiting his usual exactitude, had calculated the expected intensity of the pile. Midway through the process of drawing out the final control rod, he declared he was hungry, and the group broke for lunch. When the researchers resumed the process, instrumentation recorded the chain reaction taking place exactly as he had predicted. At that, the scientists produced a bottle of Chianti and celebrated their feat.[14]

With confirmation of the possibility of a chain reaction, and with verification that such a pile could produce fissionable plutonium, physicists had taken a major step forward. The next step entailed working with a series of processes—centrifugal, gaseous diffusion, electromagnetic, pile—to isolate U-235. In ordinary circumstances, researchers might have tried one approach; and if it proved unsuccessful, then they would have experimented with another. Facing the demands of war, with money no object, they built installations at Oak Ridge, Tennessee; Hanford, Washington; and elsewhere and embarked on all routes at once.[15]

Meanwhile, other scientists assembled at a laboratory in Los Alamos, in the New Mexico desert, where the bomb was to be built and tested. J. Robert Oppenheimer, director of the laboratory and one of the key figures in the entire Manhattan Project, had selected the site. Though he had not won a Nobel Prize, he was a brilliant physicist who could convey his passion and insight to both students and colleagues. Educated at Harvard, then at the Cavendish Laboratories in Cambridge, England, and at the University of Göttingen in Germany before returning to teach at the University of California at Berkeley, the gaunt Oppenheimer was a lively presence at social and scientific gatherings. He was, a protégé recalled, "the fastest thinker I've ever met." He was also a man of broad and sometimes eccentric interests. He read widely—novels, plays, and poems; he studied Sanskrit; he taught himself about other areas of science. Only politics failed to interest him, at least until the 1930s, when the plight of Jewish relatives in Germany and the difficulties of his students during the Depression began to disturb him. Relationships with two women—the first, his fiancée; the second, his wife—drew him into left-wing causes, until the United States entered World War II. Leslie Groves, who called Oppenheimer "a real genius," insisted on appointing him to head the new laboratory, despite security questions about his past.[16]

Aware of the enormous challenge American science faced, Oppenheimer knew he had to recruit the country's top physicists fast. First he had to convince the Office of Scientific Research and Development to release the scientists from other laboratories, where many of them were involved in the effort to create radar and similar devices for use in the war. Then he had to persuade the scientists themselves to join him in the barren desert, where they and their families would face isolation, travel restrictions, and the disagreeable prospect of military control. Some, like Emilio Segrè, were overwhelmed by the "beautiful and savage country." Others shared the less sanguine view of Leo Szilard. "Nobody," he told colleagues in Chicago, "could think straight in a place like that." Some potential recruits, like I. I. Rabi, who was working on radar at the Massachusetts Institute of Technology (MIT) turned Oppenheimer down. Other noted physicists, like Hans Bethe and Edward Teller, yielded to Oppenheimer's pleas and followed him to Los Alamos. The staff, when in place, was probably the most distinguished group of scientists ever assembled for an extended period of time.[17]

For those who went to Los Alamos, it was a remarkable experience. Rabi, who finally agreed to visit occasionally as a consultant, recalled entering "a new world, a mystic world." There was a sense of camaraderie as the world's greatest scientific minds worked and played together. Teller brought his Steinway piano; chemist George Kistiakowsky taught mathematicians John Von Neumann and Stanislaw Ulam to play poker; Oppenheimer and others put on a version of *Arsenic and Old Lace*. Paul Olum, a 24-year-old Princeton graduate student in theoretical physics and mathematics who followed his mentor to New Mexico, "learned to ski at Los Alamos, taught by the European scientists, and we went out and climbed in the mountains on skis and talked, mostly about science."[18]

The scientists trying to design "the gadget" faced a series of complex problems. They needed to determine how much fissionable material was necessary, how it would react under stress, and how it could be brought together to cause an explosive chain reaction. A gun method, firing one chunk of subcritical material into another chunk of subcritical material, seemed promising with uranium but not with plutonium. Implosion, produced by arranging conventional explosives to cause a converging shock wave and collapse a mass of plutonium to the point of criticality, appeared to be the answer.[19]

By mid-1945, scientists were ready to test a device. Confident that a uranium weapon would work, they needed to make sure a plutonium bomb would function. Oppenheimer picked a site at Alamogordo, about two hundred miles from Los Alamos, for the ultimate experiment, scheduled for July 16. He had a John Donne poem in mind when he assigned the name Trinity to the test:

> Batter my heart, three-person'd God, for you
> As yet but knock, breathe, shine, and seek to mend.
> That I may rise and stand, o'erthrow me, and bend
> Your force, to break, blow, burn, and make me new.

Science, he hoped, might transform the world.[20]

Apprehensive, the scientists contributed to a betting pool predicting the yield, applied suntan lotion for protection from the flash, and watched as the plutonium device performed exactly as planned. The scientists were overwhelmed. Otto Frisch recalled how the bomb broke the pre-dawn darkness: "And then, without a sound, the sun was shining; or so it looked." I. I. Rabi was more dramatic: "It blasted; it pounced; it bored its way into you. It was a vision which was seen with more than the eye." Oppenheimer was reminded of a passage from the *Bhagavad Gita*, the Hindu holy book, in which Vishnu took on a fiery, multi-faceted form: "Now I am become Death, the destroyer of worlds." Kenneth Bainbridge looked at Oppenheimer and said, "We're all sons of bitches now."[21]

The bombs were just about ready for use. Though scientists wanted to help determine the next steps, others, charged with ultimate responsibility for conduct of the war, insisted on deciding what to do with the weapons. Having played their part superbly and having created the new explosive as asked, the scientists found themselves on the periphery of the decision-making process. Despite their growing sense of accountability for the weapon, they had to yield to those more conversant with military and diplomatic affairs—and more eager to use the bomb against the enemy—in one of the first confrontations between scientists and government officials in the atomic age.

Harry S Truman, who inherited the presidency from Franklin Roosevelt, had to make the final decisions. Truman was a far cry from his confident predecessor, who had been elected four times to the nation's top office. An unpretentious senator from Missouri before selection as the Democratic party's vice-presidential candidate in 1944, Truman was ill-prepared to take over the White House. He had met with Roosevelt only twice in the three months before FDR's death on April 12, 1945. Insecure as he assumed power, he told a former Senate colleague, "I'm not big enough for this job," and others, unable to imagine anyone but Roosevelt as President, agreed. Encountering reporters after a first meeting with legislative leaders, he told them, "Boys, if you ever pray, pray for me now." Yet Truman learned fast and grew in office. The sign on his desk reading "The Buck Stops Here" reflected his willingness to take responsibility for decisions made, and he was known to make them quickly. But quick decisions were not necessarily the best ones, and critics often wondered if he had considered all the consequences.[22]

On assuming power, Truman knew little about the atomic bomb. Twice, in 1943 and 1944, while chairman of a Senate committee investigating national defense, he had asked for an accounting of huge expenditures. Telling him that this was "the greatest project in the history of the world" but that it was "most top secret," Secretary of War Stimson had persuaded Truman to curtail the investigation.[23]

Now, however, Truman was President and needed to know about the project. After Truman's swearing-in ceremony and the short Cabinet meeting that followed, Stimson lingered to tell Truman about "a new explosive of almost unbelievable destructive power." A week and a half later, the Secretary of War brought Truman up to date about progress on the Manhattan Project and informed him about the dangers surrounding "the most terrible weapon ever known in human history." Stimson warned of the possibility of an arms race and an atomic war. "If the problem of the proper use of this weapon can be solved," he wrote in a memorandum he presented to Truman, "we would have the opportunity to bring the world into a pattern in which the peace of the world and our civilization can be saved." To that end, he suggested that the President appoint a special committee to consider the issues raised by the atomic bomb.[24]

The Interim Committee Truman named included Stimson; James F. Byrnes, a former Supreme Court justice and Secretary of State–designate; science administrators Bush and James B. Conant (who succeeded Bush as chairman of the National Defense Research Committee), the president of MIT; and representatives of the Departments of the Navy and State. A Scientific Panel, reporting back to the Interim Committee, consisted of Oppenheimer; Fermi; Ernest Lawrence, Manhattan Project leader at Berkeley; and Arthur H. Compton, head of the Metallurgical Laboratory at the University of Chicago. Although there was no military panel, Army Chief of Staff George C. Marshall and Groves attended many of the sessions of the Interim Committee.

Charged with examining how the bomb's use would affect postwar development and control of atomic energy, the committee never specifically addressed the question of whether the weapon should be used. After reviewing the background of the Manhattan Project, reflecting on future prospects for atomic development, and considering the possible effects of the bomb on the Japanese, the committee recommended, on May 31, that the bomb be dropped as soon as possible on a military target. With Germany defeated, the atomic bomb should be used to bring the war against Japan to an end. Truman accepted the advice easily, for as he later recalled: "The conclusions of the committee were similar to my own, although I reached mine independently."[25]

Meanwhile, some of the atomic scientists had become increasingly concerned about how the new weapon might be used. Leo Szilard, the driving force who had focused attention on the military possibilities of atomic energy and so started the Manhattan Project, feared the consequences of scientific success even before the Trinity test. When Szilard was unable to gain access to Truman in May, he led a delegation to speak with James Byrnes, the President's personal representative on the Interim Committee. A memorandum he presented argued that in preparing to use atomic bombs, the United States was "moving along a road leading to the destruction of the strong position hitherto occupied in the world." Any decision had to follow from careful consideration of the situation in the future, not simply the present. In evaluating that situation, scientists had to be involved.[26]

Scientists in Chicago, organized into several committees in early June by Arthur Compton, examined the same issues that concerned Szilard. One group, headed by physicist James Franck, predicted the nearly unlimited destructive capabilities of nuclear weapons and counseled against a surprise atomic attack on Japan. Such an introduction could compromise future international agreement. When Franck tried to present his report to Stimson in Washington, he found the Secretary unavailable. Increasingly estranged from the policymakers, anxious scientists found it hard to make themselves heard.[27]

The Scientific Panel took up the question of military use that concerned the Chicago scientists. Compton, who had provided a cover letter to the Franck report, was not entirely sympathetic to the report's position. Though Fermi and Lawrence were reluctant to drop the bomb without a prior demonstration, Oppenheimer helped shape a different consensus. The Los Alamos director, Szilard once observed, "would not resist using the bomb after working so hard to give it life; Oppie had acquired a stake in displaying his weapon's terrible potency on a Japanese city." On June 16, the panel informed Stimson that it could "propose no technical demonstration likely to bring an end to the war" and it could see "no acceptable alternative to direct military use." Petitions protesting a surprise atomic attack on Japan brought no further change. Signed by scientists in Chicago and Oak Ridge but squelched by Oppenheimer in Los Alamos, the petitions never reached the President. The only flexibility came on the question of notifying American allies about the bomb's forthcoming use. The Interim Committee accepted the recommendation of the Scientific Panel that the allies be so informed.[28]

Truman was meeting with his allies at the Potsdam Conference in July 1945 when he learned of the successful Trinity test. As he faced the decision that was his alone to make, he wondered in his private diary whether the new weapon might "be the fire [of] destruction prophesied in the Euphrates Valley Era, after Noah and his fabulous ark." He also recorded his intention to tell Soviet leader Joseph Stalin about the bomb, as the Interim Committee had recommended, for he felt that it might well preclude the Soviet Union's planned entry into the war against Japan. "Believe Japs will fold up before Russia comes in," Truman noted. "I am sure they will when Manhattan appears over their homeland."

Despite his stated intention of informing Stalin, Truman engaged in a charade aimed at keeping him in the dark. Before he left Potsdam, the President sauntered over to the Soviet leader and told him that the United States had a new weapon of great explosive force. Stalin, already briefed by Russian spies, replied simply that he hoped the Americans would use the weapon against Japan.[29]

The atomic bomb preempted alternative ways of ending the war that might have been tried. Conventional military maneuvers had defeated Germany and an invasion of Japan already scheduled promised the same result. A naval blockade or continued aerial bombing might also bring Japan to the point of surrender. But the Navy did not press its position and the Army saw little reason to risk further American casualties, which could be particularly heavy in the face of fiercer Japanese resistance as the war came closer and closer to the home islands. Some diplomats believed that Japan might surrender more quickly if the United States softened its insistence on unconditional surrender and assured the Japanese that the Emperor could keep his throne, but others were less willing to compromise. After years of fighting that began with the sneak attack on Pearl Harbor, anti-Japanese sentiment remained strong. In the field, correspondent Ernie Pyle wrote, "the Japanese are looked upon as something inhuman and squirmy—like some people feel about cockroaches or mice." Back home, feelings were much the same.[30]

The bomb was attractive because it could provide a dramatic and conclusive demonstration of American might to the Japanese. It could also furnish a similar demonstration of American strength to the Russians. At a time when the Grand Alliance was beginning to splinter and Soviet–American tensions, particularly in eastern Europe, were becoming more intense, the bomb might prove useful. Reflecting on relations with the Russians, Stimson noted in his diary, "Over any such tangled wave of problems the S-1 [atomic bomb] secret would be dominant." Use of the weapon could underscore America's determination to stand firm and, equally important, prevent Soviet entrance into the Pacific war, something Stalin had promised at the Yalta Conference in February 1945.[31]

Clearly cognizant of its potential impact on both enemies and friends, Truman accepted the bomb as a legitimate, though uniquely powerful, military weapon. When faced with a final decision, he chose to use it because most scientists and policymakers had always assumed that it would be used. From the first days of the Manhattan Project, the process of creation aimed at producing a weapon of war. The atomic bomb was far more potent than conventional explosives, to be sure; but it was simply one more item in an arsenal that had already brought tremendous destruction to Germany and Japan.

The use of the bomb was a foregone conclusion. Truman would have been hard pressed to make any other decision. He inherited Roosevelt's advisers and his policies, and Roosevelt had intended to use the bomb. Indeed, Roosevelt, who sought accommodation with Stalin, accepted British Prime Minister Winston Churchill's assumption that the bomb would serve as a hedge against postwar Russian ambitions. Germany was the intended target while Roosevelt was still alive, but the European war ended soon after he died, and the struggle against Japan continued. It became the natural target for Truman, as it would have for FDR. In office barely four months when the first bombs were ready, Truman would have had to exert tremendous force to overturn a decision that had effectively been made three years before. He responded to the advice of his staff and authorized the use of the new weapon because such a decision was the obvious one given the momentum of the developmental process. As Leslie Groves later observed, Truman's "decision was one of noninterference— basically a decision not to upset the existing plans."[32]

On August 6, a B-29 dropped "Little Boy," a uranium bomb, on Hiroshima, killing seventy thousand and injuring seventy thousand more. "My God," the copilot scribbled as he watched the bomb explode. On August 9, another B-29 dropped "Fat Man," a plutonium bomb, on Nagasaki, this time killing forty thousand and injuring another sixty thousand. The United States intended to use such weapons as quickly as they could be constructed to topple Japan.[33]

The bombs devastated the Japanese cities. Hiroshima, an important military center, had been spared the raids that had destroyed other cities; American officials wanted to keep it intact to demonstrate better the force of the new bomb. On the fateful morning, as workers hurried to factories, students began classes, and children played in the streets, no one had any inkling of what lay ahead. An early air-raid alarm ended with an all clear signal. Then the *Enola Gay* sliced through the cloudless sky and plunged the city into what novelist Masuji Ibuse has called "a hell of unspeakable torments."[34]

Exploding more than five hundred meters above the ground,

the bomb shattered the calm. Ibuse, a distinguished Japanese author who was near Hiroshima on August 6, later recounted the
story through the detailed journal of Shigematsu Shizuma in *Black
Rain*: "I saw a ball of blindingly intense light, and simultaneously
I was plunged into total, unseeing darkness. The next instant, the
black veil in which I seemed to be enveloped was pierced by cries
and screams of pain, shouts of 'Get off!' and 'Let me by!,' curses,
and other voices in indescribable confusion." A mushroom cloud
followed, "shaped more like a jellyfish than a mushroom. Yet it
seemed to have a more animal vitality than any jellyfish, with its
leg that quivered and its head that changed color as it sprawled
out slowly toward the southeast, writhing and raging as though it
might hurl itself on our heads at any moment. It was an envoy of
the devil himself, I decided."[35]

The fireball, its temperature three hundred thousand degrees
centigrade at the center, shattered houses and pulverized people
within two kilometers of the hypocenter. In both Hiroshima and
Nagasaki, some of the victims evaporated in the heat; others turned
into charred corpses. Many of those who survived were horribly
burned. Ibuse described a Hiroshima resident, whose "back was
red and lumpy like a turkey's comb, and the skin had come off
like a sheet of oiled paper." Tatsue Urata, writing about Nagasaki
six years after the bomb fell there, pictured another victim: "The
skin was peeling off his face and chest and hands. He was black
all over—I suppose it was dirt that had stuck to him where the
skin had peeled off; his whole body was coated with it and the
blood trickling from his wounds made red streaks in the black."
Kimono patterns were sometimes burned onto remaining skin.[36]

Survivors in both cities faced total chaos. Some rushed to river
banks for water. Others sought parents or children in the piles of
debris, "searching for remains among the rubble," Ibuse observed,
"like people seen searching for shellfish on the beach." Babies
suckled the breasts of dead mothers; mothers clung to dead infants,
unwilling to let them go. Corpses cluttered streets and fields. They
were "swollen round," Matsu Moriuchi of Nagasaki noted, "and
looked a little like watermelons in a patch." Hospitals, like other
buildings, had been destroyed, and doctors trying to help had only
the most primitive facilities. When one official described in *Black
Rain* tried to notify his headquarters about the extent of the damage, he found that headquarters no longer existed.[37]

The Japanese were overwhelmed. Two days after the Hiro-

shima bomb, the Soviet Union entered into the war, and before
that shock had subsided, the second bomb fell on Nagasaki. The
Japanese had already been considering surrender but were unwilling to relinquish their Emperor; now they sought to extricate
themselves from the conflict in whatever way they could. On August 14, the Japanese accepted American terms, which implicitly
permitted the Emperor to retain his throne, and on September 2
the formal surrender took place.[38]

American officials were delighted with the results. Aboard
ship coming home from Potsdam when he heard about the Hiroshima bomb, Truman told the sailors around him, "This is the
greatest thing in history." He then informed them of the new
weapon's power and voiced his hope that the Pacific war was nearly
over. To the public at large, he issued a prepared statement describing the new atomic bomb as "a harnessing of the basic power
of the universe," a mastering of "the force from which the sun
draws its power." With such power in hand, he warned that unless
the Japanese capitulated, "they may expect a rain of ruin from the
air, the like of which has never been seen on this earth."[39]

Truman's enthusiasm reflected a naive American hope that
new weapons could help end war forever. He was part of a culture
that had long fantasized about ultimate weapons, and those fantasies helped shape thinking about the bomb. Like Woodrow Wilson, Truman envisioned a peaceful world rising from the wreckage
of war. In 1910, he had copied—and often carried—ten lines from
Alfred Lord Tennyson's 1842 poem "Locksley Hall":

For I dipt into the future, far as human eye could see,
Saw the Vision of the World, and all the wonder that would be; . . .

Heard the heavens fill with shouting, and there rain'd a ghastly dew
From the nations' airy navies grappling in the central blue; . . .

Till the war-drum throbb'd no longer, and the battle-flags were furl'd
In the Parliament of man, the Federation of the world.

Waiting for the results of the Trinity test while en route to Potsdam
for his first meeting with British and Soviet leaders, he had pulled
the poem from his wallet and had recited the lines for a reporter.
The bombs, dropped soon after, might help attain that dream.[40]

Enthusiasm notwithstanding, Truman recognized the formidable implications of atomic energy. "We have discovered the most terrible bomb in the history of the world," he noted after the Trinity test. The wartime use of the bombs, with such tremendous loss of life, underscored that judgment. After the war, even as he authorized expansion of the atomic arsenal, Truman occasionally wondered what role the frightful new weapon could play in world affairs. In the fall of 1945, as Truman worried about the international situation, his budget director Harold Smith assured him, "Mr. President, you have an atomic bomb up your sleeve"; Truman replied, "Yes, but I am not sure it can ever be used."[41]

Yet Truman remained convinced he had made the right decision in dropping the bomb on the Japanese. For him it remained "a powerful new weapon of war" that "had to be used to end the unnecessary slaughter on both sides," and he never shirked responsibility for mandating its use. "The final decision of where and when to use the atomic bomb was up to me," he recorded in his *Memoirs*. "Let there be no mistake about it." He "never had any doubt that it should be used" and never had second thoughts about the choice he made. Later, in 1959, he claimed that after making the decision he slept soundly. In 1965 he declared, "I could not worry about what history would say about my personal morality. I made the only decision I ever knew how to make. I did what I thought was right."[42]

When news of the Hiroshima blast became public, Americans responded as Truman had. A wave of enthusiasm swept the country, tempered periodically by a sense of foreboding about the destructive possibilities of the new weapons of war. The nation sometimes worried about the future but seldom questioned the decision to drop the bomb.

The news came as a tremendous surprise. Because of the top-secret status of the Manhattan Project, Americans knew little about the effort to create a bomb. When Cleve Cartmill published the story "Deadline" in *Astounding Science Fiction* in March 1944, agents of the Federal Bureau of Investigation (FBI) were concerned about his remarkably accurate description of a U-235 fission weapon and attempted to persuade his editor to cease such publication. Though that effort was unsuccessful, the public remained largely unaware of the development of the bombs until they were dropped on Japan.[43]

Then Americans learned a great deal about the experimental work done during the war and about the powerful weapon it had produced. Yet the news that reached the public was only what government officials chose to reveal. In the first effort to orchestrate opinion during the atomic age, policymakers carefully shaped public perceptions of the new weapon as they highlighted the bomb's role in ending the brutal war and the possibilities it offered for a brave new nuclear world. Once again, as when they made the decision about how to use the bomb, they seized the initiative from the scientists who had created the weapon. The War Department released a number of reports written earlier by *New York Times* science reporter William L. Laurence, who served as Manhattan Project publicist. On loan from the newspaper, his whereabouts known only by his managing editor and his wife, he provided the background material for media stories that appeared as soon as the bombs were dropped and also published a number of personal accounts. In *Life* magazine several weeks after the end of the war, Laurence described the "giant flash" of the Nagasaki bomb turning into "bluish-green light that illuminated the entire sky." He pictured the fireball for readers, then noted the transformation of the cloud: "As the first mushroom floated off into the blue, it changed its shape into a flowerlike form, its giant petal curving downward, creamy-white outside, rose-colored inside. It still retained that shape when we last gazed at it from a distance of about 200 miles." In the months that followed, Laurence continued to write, sometimes with the same millennial imagery, of past development and of prospects ahead. Noting the timeless "rebellion against the limitations of space and time" in the spring of 1946, he suggested that scientists had found "the philosopher's stone that will not only transmute the elements and create wealth far greater in value than gold but will also provide him with the means for gaining a far deeper insight into the mysteries of living processes, leading to the postponement of old age and the prolongation of life." Other commentaries pictured the apocalyptic power of the new weapon. A Paramount newsreel declared that Hiroshima was obliterated by a "cosmic power . . . hell-fire . . . described by eyewitnesses as Doomsday itself!"[44]

It was not hard to persuade Americans of the bomb's value. Soldiers—and their families back home—felt an overwhelming sense of relief that the amazing weapon had brought a quick end to the war. Literary critic Paul Fussell has recently recalled the brutality of the struggle and the terror of "having to come to grips, face to face, with an enemy who designs your death." Like countless other soldiers, author William Manchester among them, Fussell feared the forthcoming invasion with its expected loss of life. Once nuclear weapons had secured Japanese surrender, he could only say, "Thank God for the Atom Bomb." The music industry seized upon the same theme. "When the Atom Bomb Fell," recorded in December 1945, was probably the first musical reaction to the atomic blasts. Recorded by Karl Davis and Harty Taylor, country and western singers from rural Kentucky working in Chicago, the song reflected a combination of awe and relief. The second verse had lyrics of vivid imagery describing the extraordinary power of the blast and closed with a repeated refrain of military thanks:

> Smoke and fire it did flow through the land of Tokyo.
> There was brimstone and dust everywhere.
> When it all cleared away, there the cruel Japs did lay.
> The answer to our fighting boys' prayer.

Terrible though the carnage may have been, it was justified in the end for having saved American lives.[45]

Early public reactions to Hiroshima and Nagasaki were often light-hearted and glib. Hours after the announcement of the first bomb, the Washington Press Club prepared an "Atomic Cocktail," made from a combination of Pernod and gin. Los Angeles burlesque houses featured "Atombomb Dancers." Jokes, sometimes antidotes to stress, played upon atomic themes. One commentator observed that Hiroshima "looked like Ebbets Field after a game between the Giants and the Dodgers." Another declared, "They should call that bomb Up and Atom; when it blew up and the smoke cleared away, the Russians were at 'em." Still another quipped that Japan suffered from "atomic ache." *Life* magazine noted, with a full-page photograph of a starlet in a two-piece bathing suit, that on August 8 Linda Christians had been named "Miss Anatomic Bomb."[46]

The light-hearted mood persisted as the atom became an integral part of American life. Several years after the first blast, a Boston hotel still served a rich dessert—consisting of champagne-soaked sherbet spiced with liqueur—known as an "Atomic Bomb." In 1946, the General Mills Corporation provided children with an

"Atomic 'Bomb' Ring" for fifteen cents and a Kix cereal box top. By looking into the "sealed atomic chamber," the wearer could "see genuine atoms split to smithereens!" About three quarters of a million children ordered rings. The public vocabulary absorbed atomic terms. Following the first postwar atomic test in the Pacific, a French fashion designer called a new and skimpy bathing suit a "Bikini," to draw attention to its explosive promise. The song "Atomic Power," written the day after Hiroshima by cowboy singer Fred Kirby and recorded six months later by the Buchanan Brothers, first noted the brimstone fire from heaven, then struck a religious theme in the chorus: "Atomic power, atomic power, was given by the mighty hand of God." It soon appeared on *Billboard* charts and remained popular for the next twenty years.[47]

At times, atomic possibilities seemed boundless. David E. Lilienthal, first chairman of the Atomic Energy Commission, recalled that the atom "had us bewitched. It was so gigantic, so terrible, so beyond the power of imagination to embrace, that it seemed to be the ultimate fact. It would either destroy us all or it would bring about the millennium. It was the final secret of Nature, greater by far than man himself, and it was, it seemed, invulnerable to the ordinary processes of life, the processes of growth, decay, change." Some Americans assumed that the atom could do anything. A Newport, Arkansas, farmer wrote to atomic authorities in Oak Ridge, Tennessee, with a curious request: "I have some stumps in my field that I should like to blow out. Have you got any atomic bombs the right size for the job? If you have let me know by return mail, and let me know how much they cost. I think I should like them better than dynamite." When asked about possible atomic applications, Enrico Fermi once remarked, only half in jest, "It would be nice if it could cure the common cold." Contemplating future applications even before the second bomb was dropped on Nagasaki, the *New York Times* speculated about an atomic plane. Similarly, a spaceship might now have the power to reach the Moon or Mars. Cars might run on tiny pellets of atomic fuel, and power, of course, would be plentiful and cheap. William Laurence was but one of those who helped encourage the myth of a boundless nuclear future. While discounting the most outlandish misconceptions, he still wrote glowingly about "the gateway to a new world."[48]

Enchanted by the bomb and the future it foreshadowed,

Americans were sure they had acted wisely in developing and dropping the new weapon. A Gallup poll in September 1945 indicated that 69 percent of the public felt it was "a good thing" the bomb had been constructed. When asked about wartime use, the results were much the same. Two weeks after Hiroshima, a Gallup survey revealed that 85 percent of the respondents supported the bombing. Two months after the Japanese surrender, a *Fortune* poll showed that while many Americans now felt remorse, 53.5 percent still endorsed "without reservation" what had been done. "Not the least extraordinary fact about the postwar period," philosopher Lewis Mumford noted at the time, "is that mass extermination has awakened so little moral protest."[49]

Despite such support, a number of nationally known commentators recognized the more ominous implications of the bomb from the very start. Some of the nation's most popular radio broadcasters voiced their personal fears as they reacted to what had been done at Hiroshima. "For all we know, we have created a Frankenstein!" declared H. V. Kaltenborn in one of the first public comments on the bomb. Edward R. Murrow observed a few days later, "Seldom, if ever, has a war ended leaving the victors with such a sense of uncertainty and fear, with such a realization that the future is obscure and that survival is not assured." Writers made much the same point. Destruction might prove uncontrollable, playwright George Bernard Shaw warned in the *Washington Post* in mid-August; the bomb might "burn down a house to roast a pig." Apprehension was unavoidable as the implications of the extraordinary weapon became visible. "Whatever elation there is in the world today," editor Norman Cousins wrote in *The Saturday Review*, "is severely tempered by fear. . . . a primitive fear, the fear of the unknown, the fear of forces man can neither channel nor comprehend." The fear was not new, "but overnight it has become intensified, magnified." Author E. B. White observed similar rumblings in *The New Yorker* as the war wound down: "For the first time in our lives, we can feel the disturbing vibrations of complete human readjustment. Usually the vibrations are so faint as to go unnoticed. This time, they are so strong that even the ending of a war is overshadowed." Now, White added, we have to deal with the spectacle of man "stealing God's stuff."[50]

In communicating their own fears, authors and announcers developed a more widespread constituency of concern. As they responded, some Americans echoed anxieties scientists felt as they worked furiously to develop the bomb. Waiting for the Trinity test, Enrico Fermi had offered to take bets on whether the device would ignite the atmosphere, and if so, whether it would wipe out just New Mexico or the entire world. Despite relief at the Alamogordo result, fears persisted after Hiroshima and Nagasaki that even larger bombs might produce such a chain reaction in the atmosphere, the earth, or the sea. In early 1946, Hans Bethe finally eased those anxieties with calculations proving that there was "no danger of a nuclear explosion in any substance naturally occurring on earth with any of the atomic bombs which have been developed or have been conceived on paper."[51]

Other Americans feared the results of a future war as they became more aware of the magnitude of the new weapon. In September 1945, only 37 percent of the respondents to a National Opinion Research Center poll perceived a real danger of an atomic bomb being dropped on the United States in the next twenty-five years. A different study in the summer of 1946—this one conducted by Leonard S. Cottrell, Jr., and Sylvia Eberhart for the Social Science Research Council—found that 64 percent of those polled believed that bombs would someday be used against the United States. Other polls revealed serious apprehensions about the possibility of another world war. And if such a conflict occurred, 83 percent of one early sample believed there was a real danger that "most city people on earth" would be killed.[52]

Fears became more pronounced as the entire literate public learned more about the ravages of the bomb. The first accounts were antiseptic and spared the most gruesome details. The earliest photographs were grainy images of columns of smoke, with little reference to human destruction. *Life* magazine, on August 20, 1945, ran a lengthy spread which began with an artistic rendition of the Hiroshima bomb showing clouds of billowing smoke obscuring the barely visible city below. Real photographs several pages later captured Hiroshima before and after, but the leveled landscape seemed no different from accompanying pictures of the devastation caused by scores of conventional bombs. While newsreels and photographs soon provided better images of the desolate wasteland, they seldom focused on people.[53]

In a pre-television age, before Americans came to expect grisly images on the nightly news, it fell to the world of print journalism to convey the human suffering caused by the bomb. John Hersey, 31-year-old author of the already popular novel A *Bell for Adano* and other books and stories stemming from his experiences in the war, visited Japan and wrote about the situation in Hiroshima just after the bomb exploded. His account, published in the August 31, 1946, issue of *The New Yorker*, overwhelmed readers. Newspapers ran the entire text, and radio stations read it aloud. The book version, distributed free to many Book-of-the-Month-Club members, became a best-seller and enjoyed the vast popular audience that comes to but an occasional book.[54]

Hiroshima offered dispassionate but detailed descriptions of six residents of Hiroshima: an industrial clerk, a seamstress, a young physician, an older physician, a Japanese Methodist minister, and a German priest. Hersey wrote of their routine daily activities on the morning of August 6, then recorded their actions as the bomb fell. From time to time he included scenes of the horrible suffering they encountered, as when he described the hundreds of people fleeing the city: "The eyebrows of some were burned off and skin hung from their faces and hands. . . . Some were vomiting as they walked." Sometimes he was even more graphic, writing of the mouths of maimed soldiers as "mere swollen, pus-covered wounds." But for the most part, the account was understated, hardly sentimental or melodramatic. Hersey deliberately wrote sparely, unemotionally, of the mundane reactions of his protagonists as they struggled to cope with a crisis they only barely understood, and so helped accentuate the horror of the fateful day.[55]

Hiroshima left readers stunned. It gave them a sense of what the Japanese had experienced, a framework within which to incorporate images of the bomb. Many expressed anxiety after reading the account; some who finished it late at night reported fitful sleep and disturbing dreams. Critics applauded the essay. "Nothing that can be said about the book can equal what the book has to say," reviewer Charles Poore declared in the *New York Times*. "It speaks for itself, and, in an unforgettable way, for humanity." Anthropologist Ruth Benedict hailed the work's "stark simplicity" in *The Nation* and suggested that "the calmness of the narrative throws into relief the nightmare magnitude of the destructive power the brains of man have brought into being." More fully aware now of what the weapon had done, readers feared what a future bomb could do.[56]

In the first year of the atomic age, many Americans asked whether the new bomb was any different from other weapons of war. Conventional attacks on German cities had already wrought horrible destruction. In the summer of 1943, a raid on Hamburg by 731 bombers ignited fires that trapped victims and incinerated or asphyxiated forty thousand victims. In the winter of 1945, two raids on Dresden caused a catastrophic firestorm that killed thirty-five thousand people. Japanese cities suffered much the same fate. On March 9, 1945, B-29s sowed Tokyo with napalm- and magnesium-filled bombs that created an even worse inferno than that in Dresden. Rather than sweeping toward the center, the firestorm rushed outward, engulfing the entire city. Trying to flee, thousands found bridges destroyed and escape routes in flames. They collapsed, as one survivor recalled, "like so many fish left gasping in the bottom of a lake that has been drained." Sixteen square miles of the city were entirely burned to the ground; eighty-four thousand people died. This was probably the worst human-caused catastrophe of the war. What, then, made the bombs dropped on Hiroshima and Nagasaki so much worse?[57]

Numerous commentators suggested that there was little difference at all. Physicist Philip Morrison later described the position of most of the Los Alamos scientists: "From our point of view, the atomic bomb was not a discontinuity. We were just carrying on more of the same, only it was much cheaper. . . . We had already destroyed sixty-six [cities]; what's two more?" On the day of the Nagasaki bomb, the *Chicago Sun* declared: "There is no scale of values which makes a TNT explosion right and a uranium explosion wrong." Early the next year, Major Alexander de Seversky, author of a volume about air power, argued in *The Reader's Digest* that atomic devastation "had been wildly exaggerated," that "Hiroshima looked exactly like all the other burned-out cities in Japan."[58]

Yet there was a wave of revulsion, centered in religious circles, suggesting that the new bombs were different and that the moral implications of atomic energy could not be ignored. Soon after Hiroshima, thirty-four Protestant clergymen took such a stance and condemned the decision to drop the bomb in a letter to Harry

Truman. The interdenominational *Christian Century* said in mid-August that "short of blowing up the planet, this is the ultimate in violence." It was another "technical development in the art of killing" which "may lead to the extermination of man." Though the magazine tempered its position in subsequent weeks, it had made an important point that echoed in later years; the Hiroshima and Nagasaki bombs had not destroyed their targets incrementally but in one brief and awful moment. Similar destruction by even more powerful bombs was possible at any time. Though the ultimate damage was no different and the rubble looked the same, the stakes had been appreciably raised.[59]

In those first months after Hiroshima, most Americans sensed that a new era had begun. Almost without exception, they knew about the bomb; a 1946 survey by the Social Science Research Council reported that 98 percent of the population was aware of what it had done. Its dramatic and near-universal impact on public consciousness ensured its influence in all areas of American life.[60]

As the nation began to consider the dictates of nuclear strategy and the dreams of atomic development, it found itself following patterns established in the war years. Scientists working on the Manhattan Project had struggled with their own qualms but had done all that was asked in the interest of victory in a gruesome war. They had fought successfully with government officials—military and civilian—over how scientific work should be conducted and how the results should be shared. When they sought to influence the decision about the bomb's use, they soon realized the limited audience for their advice. The President and his advisers seized the initiative and made the crucial decisions with but scant attention to the concerns of the builders of the bomb. Then the administration went one step further by orchestrating the release of information and so shaping public perceptions of the emerging patterns of the atomic age. That effort, however, was not wholly successful, for fears, present from the start, could never wholly be contained and surfaced as the public learned more about the bomb. As the new weapon's potent force and future prospects became increasingly visible, critics made their voices heard in what became an ongoing and often heated nuclear dialogue. Whether they were participants or observers, most Americans understood that their world would never be the same.

2

The Question
of Control

Even before the first bombs fell, some Americans recognized the need to bring atomic energy under effective control. The Manhattan Project scientists who questioned the new weapon's wartime use raised issues that reverberated through the immediate postwar years, as the nation considered how the new force should be handled, both within the United States and in the world at large. Should the nuclear secret be shared, and if so, with whom? Could sharing American expertise head off a disastrous arms race, or would already intense international suspicion preclude any lasting agreement? And who should be responsible for making judgments about atomic development at home? Undeterred by their failure to influence nuclear strategy during the war, many atomic scientists continued to speak out and promoted the idea of international control of nuclear energy as soon as the military struggle ended. They envisioned a collaborative approach, in which the major powers would pool their knowledge and cooperate in stopping a weapons buildup before it began. Many of these scientists remained unsettled as they considered more fully the implications of the extraordinary weapon they had created and realized the need to prevent its further use. Venerated for their achievement, they shared their expertise, and their sense of foreboding, with government officials and with the public at large as they pressed for international arrangements that they insisted were necessary to ensure that no future bombs would be used. For the first year after the war, they played a major role in the nuclear dialogue and helped focus national attention on the issue. Then, as Cold War tensions became more severe, they watched policymakers retreat from the possibility of cooperation with the Soviet Union and assert the primacy of American prerogatives in the nuclear realm. That intransigence only fanned an arms race they had long feared and defined the framework of suspicion that dominated the postwar period.

All questions of nuclear control had to be addressed in the context of the Cold War. The Grand Alliance, consisting of the United States, the Soviet Union, and Great Britain, was fraught with friction throughout World War II and fragmented once the military battles were won. The breakdown in relations with the Soviet Union that occurred in the first year after the struggle drew Americans and Britons closer together. Friends during the war, partners in the effort to develop a workable bomb, they became even more reliant on one another as they hunkered down on their side of what former Prime Minister Winston Churchill called the "iron curtain." American decisions about how to handle the atomic bomb were never divorced from concern about the impact on the Russian adversary in the east.[1]

Atomic scientists in the United States were quick to demand some sort of international system of nuclear control. Many of them reacted to news of Hiroshima with elation that soon turned to gloom. Working under tremendous pressure, struggling to produce a workable weapon of war, they had succeeded in putting together all the pieces of a monumental puzzle just as they had been charged to do. Then the sense of tragedy and horror hit them. "In the summer of 1945," biochemist Eugene Rabinowitch later recalled, "some of us walked the streets of Chicago vividly imagining the sky suddenly lit by a giant fireball, the steel skeletons of skyscrapers bending into grotesque shapes and their masonry raining into the streets below, until a great cloud of dust rose and settled over the crumbling city." Some felt guilt; others felt regret. Virtually all feared the future they had helped make possible.[2]

Out of that sense of fear came a scientists' movement that

involved several thousand members of the profession. Abandoning their customary discomfort with political activism, a good number of them spoke out with missionary zeal, arguing that they had special insight that had to be taken into account. They were the ones who knew best how nuclear energy worked and what it might do. Drawing on the prestige they enjoyed for their role in creating the bomb, they now sounded a warning for the country to act responsibly to harness the atom before it was too late.[3]

Albert Einstein, whose signature on a letter to President Roosevelt had helped initiate the Manhattan Project, was one of those who felt the burdens of the new age. "I made one great mistake in my life—when I signed the letter to President Roosevelt recommending that an atomic bomb be made," he told chemist Linus Pauling not long before he died. In 1945, and in the years that followed, he underscored the need to control "the most revolutionary force since prehistoric man's discovery of fire." Scientists, he suggested, had a role to play in the process. He likened them to Alfred Nobel, the inventor of dynamite and other explosives, who later sought to atone for his discovery. "It has never been our wish to meddle in politics," he said. "But we know a few things that the politicians do not know." Einstein argued that the bomb might "intimidate the human race into bringing order into its international affairs, which, without the pressure of fear, it would not do." Specifically, he sought some kind of a world federation; it was "unthinkable that we can achieve peace without a genuine supranational organization to govern international relations." Einstein was repeatedly disappointed, both by continued nationalistic fervor around the world and by a corresponding unwillingness to heed the catastrophic possibilities of the atom. He deplored the fact that "the public, having been warned of the horrible nature of atomic warfare, has done nothing about it, and to a large extent has dismissed the warning from its consciousness." "The unleashed power of the atom," he suggested, "has changed everything save our modes of thinking, and thus we drift toward unparalleled catastrophe."[4]

Other scientists were equally concerned. Edward Teller, the talented Hungarian *émigré* who had already begun to consider the possibilities of fusion while working on fission at Los Alamos, was initially skeptical of playing a visible public role. "The accident that we worked out this dreadful thing should not give us the responsibility of having a voice in how it is to be used," he told Leo Szilard before the Trinity test. After the war, he shifted course somewhat as he described the extraordinary power of the weapons he had helped create. "We have succeeded in constructing the most spectacular machine of destruction in human history," he wrote in 1947. "There arises the possibility of a future war in which each combatant can wipe out his opponent," in which each side could do "irreparable damage to the other party." Teller implied, at this point at least, that science had to find a means to contain such destructive prospects. Later, he remained willing to speak out aggressively, even though his position shifted dramatically and his support for postwar nuclear development pitted him against other scientists committed to the need for control.[5]

Leo Szilard, active in mobilizing scientists during the war, continued to agitate in the postwar years. Initially, he tried to cooperate with policymakers as the war came to an end. "For some four to six weeks after Hiroshima," he told an audience in December 1945, "atomic scientists expressed no opinion on the political implications of the bomb, having been requested by the War Department to exercise the greatest possible reserve." He and others assumed that international discussions were taking place and wanted to lend their support to any such effort. That conciliatory spirit vanished as Szilard became embroiled in a dispute with the government over patent rights to his atomic energy discoveries. Terminated from the Manhattan Project in mid-1946, denied an Army citation by General Groves, Szilard remained an articulate outsider, convinced that scientists, having created the means of destruction, bore responsibility for the survival of the human race.[6]

Robert Oppenheimer, the nation's most visible scientist after the war, spoke out as well, though he worked as the consummate insider. Like his colleagues, he worried about the awesome destructive possibilities of the bomb he had helped create, even as he accepted the need for continued development in the atomic realm. His success in spearheading the effort to make workable weapons gave him special acclaim; virtually all scientists and policymakers looked to him for advice. Leaving California to become director of the prestigious Institute for Advanced Study at Princeton, he retained his remarkable ability to cut to the heart of any issue. As Hans Bethe later recalled of this period, "It was forever astonishing how quickly he could absorb new ideas and single out the most important point." Serving on a variety of government panels and boards, he became, in Alice Kimball Smith and Charles Weiner's phrase, "a kind of Pooh-Bah of atomic energy," active in all areas as he summarized endless arguments, wrote countless reports, and gained a hearing for his concerns.[7]

On occasion, the burden of the bomb weighed heavily on Oppenheimer. Several months after the war, he approached Truman, who wrote that the scientist "spent most of his time ringing [*sic*] his hands and telling me they had blood on them because of the discovery of atomic energy." Oppenheimer also declared a few years later, "In some sort of crude sense which no vulgarity, no humor, no over-statement can quite extinguish, the physicists have known sin, and this is a knowledge which they cannot lose." Though Ernest Lawrence, his former Berkeley colleague, responded with the comment, "I am a physicist and I have no knowledge to lose in which physics has caused me to know sin," Oppenheimer remained concerned with the explosive implications of nuclear energy, even as he worked with administration authorities on issues of atomic development.[8]

Oppenheimer understood the dilemma the scientists faced. In an often-recalled speech to the Association of Los Alamos Scientists on November 2, 1945, he shared his anxiety "about the fix we are in." The scientists had behaved appropriately during the war: "If you are a scientist you believe that it is good to find out how the world works; that it is good to find out what the realities are; that it is good to turn over to mankind at large the greatest possible power to control the world and to deal with it according to its lights and its values." But now the power released had proven far more potent than expected, the world was entirely changed, and the scientists had to deal with the consequences of their actions.[9]

In a series of articles and addresses, Oppenheimer sought to keep attention focused on the horror of nuclear war. "It did not take atomic weapons to make man want peace," he declared. "But the atomic bomb was the turn of the screw. It made the prospect of future war unendurable." Trying "to rub the edges off this new terror" was not the answer. It was important instead to return "insistently to the magnitude of the peril," for "I see in that our one great hope." Working with the government, he hoped to gain a hearing for scientific concerns and to push for a flexible policy of international cooperation and control.[10]

The atomic scientists were successful in making themselves heard in the first postwar years. Embarking upon a tremendous educational task, the alumni of Los Alamos gave lectures, appeared on radio programs, spoke to schools and service organizations, and wrote for the popular press. As early as the fall of 1945, many of them descended on Washington, D.C., to persuade members of Congress of the importance of international control. When asked for specifics, they returned with the detailed information requested and organized a lecture series on nuclear physics that sixty congressmen attended.[11]

They also formed a national organization. In November 1945, a number of them met in Washington and established the Federation of Atomic Scientists (later the Federation of American Scientists). Their fourth-floor walk-up office included, according to one of the founders, "a desk, telephone, ancient and noisy typewriter, an inadequate number of chairs, the *World Almanac*, a telephone book, and $20.00 worth of newly purchased stationary supplies." William Higinbotham, a 35-year-old physicist who had served at Los Alamos, led the group. Though not one of the nation's most celebrated scientists, he was, according to news commentator Raymond Gram Swing, "quiet, modest, lucid and impellingly convincing." The first newsletter, consisting of typed carbons, appeared days after the federation was formed. Four months later, the organization published *One World or None*, a slender paperback volume that outlined atomic hazards, described a hypothetical attack, and considered, though not entirely successfully, what might be done. With a hundred thousand copies sold, the book was a best-seller and gave the federation useful exposure.[12]

Meanwhile, the Chicago scientists launched their own publication. Informal conversations among Eugene Rabinowitch, physicist Hyman Goldsmith, sociologist Edward Shils, and others continued the dialogue begun during the war. When University of Chicago chancellor Robert Hutchins provided ten thousand dollars, a new journal became possible, and in December 1945 the inaugural issue of the *Bulletin of the Atomic Scientists* appeared. The initial board of sponsors, which included Albert Einstein, Robert Oppenheimer, and Edward Teller, provided instant prestige. Rabinowitch, the first editor-in-chief, became the journal's guiding spirit. A Russian *émigré* who had worked on the Manhattan Project and had contributed to the Franck report advising against a surprise atomic attack on Japan, he was committed, he said in 1946, to "fight to prevent science from becoming an executioner of mankind." Gentle, tireless, and optimistic, he spent his time raising money; writing about science, technology, philosophy, and politics; and keeping the journal afloat. From six hundred fifty copies in 1945, circulation rose to ten thousand the next year, as the *Bulletin* reached out to scientists and non-scientists alike in the effort to prevent nuclear disaster.[13]

The *Bulletin's* symbol became the "doomsday clock" which first appeared on the cover in June 1947. Set at seven minutes to midnight, it reflected the danger of continued nuclear testing and Cold War confrontation. Initially, the editors intended the clock to be set permanently at a fixed time; several years later, they decided it should move forward or back according to the changing threat of nuclear war. "The *Bulletin's* clock," Rabinowitch wrote, "is intended to reflect basic changes in the level of continuous danger in which mankind lives in the nuclear age, and will continue living, until society adjusts its basic attitudes and institutions to the challenge of science."[14]

Working on a variety of fronts, the scientists transformed their image. Merle Miller, one of the editors of *Harper's Magazine*, remarked in 1947 on the shift that had taken place. "Until August 1945," he wrote, "most of us laymen thought of scientists as colorless little men who wore grimy smocks and labored long hours in musty laboratories, speaking a language too technical and far too uninteresting for us to bother to understand." Hiroshima and Nagasaki changed all that. Since the war, "the scientists emerged from their laboratories and began talking, loud and long." Most important, they made sense, "because we realize that they have nothing to sell except possibly a formula for the continued existence of the human race." Physical scientists were "the vogue these days," according to a *Harper's* contributor just after the war. "No dinner party is a success without at least one physicist to explain . . . the nature of the new age in which we live."[15]

The scientists, like the Tralfamadorians—creatures from outer space Billy Pilgrim imagines as he travels through time in Kurt Vonnegut's novel *Slaughterhouse-Five*—foresaw a catastrophic future in which the world might be destroyed. They felt a special responsibility to alert others to the danger of a deadly arms race and to the need for new ways of dealing with questions of defense. As they spoke out to public and private constituencies, they drew on arguments first advanced by Danish physicist Niels Bohr during the war.[16]

Like Einstein, Bohr had changed the nature of twentieth-century science. Drawing on speculation that energy traveled in quantum—or particle—form, he applied the notion to the atom itself. His quantum theory, with its accompanying model of the atom, had won him the Nobel Prize in 1922 and established him in the forefront of nuclear physics. His research institute in Denmark drew scientists from all over the world, who respected both his accomplishments in the field of physics and his warm sense of humanity. Physically imposing, Bohr often appeared cautious and hesitant. "It didn't help that he spoke with a soft voice," author C. P. Snow observed, "not much above a whisper." Still, when he spoke, people listened. Einstein admired him for "uttering his opinions like one perpetually groping and never like one . . . in the possession of definite truth."[17]

Bohr had begun to contemplate the role of atomic energy in world affairs as he became involved with the Manhattan Project. Escaping from Nazi-occupied Denmark in 1943, he went first to England, then to the United States, where he settled in Los Alamos. Even though the world's best-known physicists passed through the area regularly, Bohr's arrival caused special excitement. His pseudonym (for purposes of security) of Nicholas Baker quickly gave way to Uncle Nick. Bohr, Oppenheimer later recalled, "made the enterprise, which often looked so macabre, seem hopeful." He shared his conviction that "the outcome would be good, and that in this the role of objectivity, friendliness, cooperation, incarnate in science, would play a helpful part; all this was something we wished very much to believe."[18]

Bohr was one of the very first scientists to consider the possible impact of the new bomb on Soviet–American relations. He understood the frictions that plagued the Grand Alliance during the war and culminated in the Cold War once the shooting had stopped. The atomic bomb, he hoped from the middle of the war on, could help ease these tensions if properly used. While lending his expertise to the enormous endeavor to create a nuclear weapon, Bohr worried about the world the scientists were helping to shape. The discovery of nuclear fission, he understood, brought new possibilities as well as new demands. "Knowledge is itself the basis of civilization," he once wrote, but "any widening of the borders of our knowledge imposes an increased responsibility on individuals and nations through the possibilities it gives for shaping the conditions of human life." Recognizing that the atomic bomb would revolutionize society, he hoped that scientists and statesmen could use it to create a more cooperative international order. Specifically, Bohr wanted to head off a destructive arms race by sharing the secret of the Manhattan Project with the Soviet Union before the bomb was ready and so involving the Russians from the start with questions of future control. He wanted to guarantee that atomic energy was "used to the benefit of all humanity and does not become a menace to civilization."[19]

Midway through the war, Bohr had sought to persuade statesmen that they could use the bomb to promote a more peaceful world order. He had spoken quietly of his ideas to other scientists at Los Alamos but really had a different audience in mind and so had approached men he thought could act on his plans. An optimist who believed that statesmen could make a difference, he naively expected national leaders to respond rationally to his arguments. He was disappointed. A meeting with Winston Churchill in May 1944 proved abortive, as the Prime Minister became bored and steered the conversation in other directions. A session with Franklin Roosevelt was more pleasant, as the President characteristically listened attentively and encouraged Bohr to think that he held the same views. But later, Roosevelt accepted Churchill's position that the secret should not be shared and ignored Bohr's plea.[20]

While Bohr's arguments fell on deaf ears during the war, they provided other scientists with a framework in the postwar years. Though the bomb had been dropped without warning on the Japanese and without adequate disclosure to the Soviets, it was not too late to arrange for a rational system of international control. Indeed, as it became clear that the Russians were intent on catching up by developing their own bomb, a cooperative order had to be established before it was too late.

Yet the scientists had to contend with others in authority who had their own ideas about the course the United States should take. Once again they found themselves engaged in a dialogue with the nation's policymakers and once again they came off second best.

Harry Truman, well aware of the awesome implications of atomic energy, had his own clear sense of American responsibility in the nuclear age. "We must constitute ourselves trustees of this new force—to prevent its misuse, and to turn it into the channels of service to mankind," he declared three days after Hiroshima. "It is an awful responsibility which has come to us. We thank God that it has come to us, instead of our enemies; and we pray that He may guide us to use it in His ways and for His purposes." Two months later, on October 3, he amplified his ideas about atomic energy in a lengthy message to Congress. Echoing Bohr, he argued that "the release of atomic energy constitutes a force too revolutionary to consider in the framework of old ideas." He also proposed discussions, first with Britain and Canada, then with other nations, to promote cooperation and "to work out arrangements covering the terms under which international collaboration and exchange of scientific information might safely proceed." But as he sought to define the shape of a new framework, he found himself less sanguine than Bohr about the Soviet Union and preoccupied instead by past assumptions about Russia's role on the world stage.[21]

Like many Americans, Truman had long been uncomfortable with communism and distressed at the Soviet purges of the 1930s. After Germany invaded Russia in June 1941, he had remarked casually to a reporter, "If we see that Germany is winning we ought to help Russia and if Russia is winning we ought to help Germany and that way let them kill as many as possible." During the war, he supported American foreign policy, which involved aiding the Soviet Union in the struggle against Hitler, but as the conflict drew to an end, he contributed to the contention with the Soviet Union that resulted in the Cold War. He believed fervently that "unless Russia is faced with an iron fist and strong language another war is in the making." Less than two weeks after he assumed the presidency, he showed his willingness to follow a hard-line policy as he called in Soviet Foreign Minister Vyacheslav Molotov and lectured him about the need for the Russians to carry out the agreements made at Yalta. "I have never been talked to like that in my life," Molotov said. "Carry out your agreements and you won't get talked to like that," Truman retorted. Relations went downhill from there.[22]

As he considered atomic issues and their impact on deteriorating relations with the Soviet Union, Truman received conflicting advice. Secretary of War Stimson, old and tired in his last stint of government service, thought long and hard about the bomb as he prepared to leave office. He could, Felix Frankfurter once noted, focus on a single thing like an old victrola needle caught in a record groove. Now he focused on ways the bomb could promote better ties with the Russians. Earlier, before the Alamogordo test, as he had contemplated diplomatic questions in eastern Europe and in the Far East, Stimson had observed in his diary, "It seems a terrible thing to gamble with such big stakes in diplomacy without having your master card in your hand." Later, after the bombs had been dropped, he had second thoughts about using the new weapon to pressure the Russians to accept American demands. He still considered "the problem of our satisfactory relations with Russia as not merely connected with but as virtually dominated by the problem of the atomic bomb." Nonetheless, it was pointless to try to cajole the Soviets while "having this weapon rather ostentatiously on our hip," for "their suspicions and their distrust of our purposes and motives will increase." Philosophically, Stimson reflected that "the chief lesson I have learned in a long life is that the only way you can make a man trustworthy is to trust him; and the surest way to make him untrustworthy is to distrust him and show your distrust." He therefore argued, like Bohr, that the United States needed to try to work cooperatively with the Soviets, for the alternative was "a secret armament race of a rather desperate character." Stimson drafted a memorandum for the President and presented his ideas at his last Cabinet meeting, but he was exhausted after a series of farewells, and, as Under Secretary of State Dean Acheson noted, "The discussion was unworthy of the subject."[23]

Stimson had some support. Secretary of Commerce Henry A. Wallace favored sharing the secret as a first step in demonstrating good faith toward the Soviet Union. Robert P. Patterson, the new Secretary of War agreed with his former chief that the Russians should be approached. Dean Acheson was even more outspoken.

"This is not a secret which we can keep to ourselves," he argued. The joint effort on the part of the British and Americans "must appear to the Soviet Union to be unanswerable evidence of an Anglo-American combination against them." For Americans "to declare ourselves trustee of the development for the benefit of the world will mean nothing more to the Russian mind than an outright policy of exclusion."[24]

Others in the administration counseled a different approach toward the Soviet Union. Like Stimson, Secretary of State Byrnes recognized the diplomatic value of the bomb, but he was less willing to share the secret. In mid-August, as the war wound down, he maintained that international arrangements were difficult to work out at this time and argued that the Manhattan Project should continue to push ahead. Citing several instances of Soviet "perfidy" early in September, he told Stimson that "we could not rely on anything in the way of promises from them." As Byrnes prepared to leave for London for a postwar meeting of foreign ministers, Stimson noted that "Byrnes was very much against any attempt to cooperate with Russia. . . . and he looks to having the presence of the bomb in his pocket, so to speak, as a great weapon to get through the thing."[25]

Military officials also wished to maintain the American monopoly. In his *Memoirs,* Truman noted that the Joint Chiefs of Staff believed that the United States should retain "all existing secrets with respect to the atomic weapons." While they spoke out in favor of political controls of some sort, they were more concerned with keeping the secret intact.[26]

The majority of the American public similarly thought it best to maintain secrecy. A Gallup poll in September 1945 revealed that 73 percent of the respondents wanted the United States to retain control of the bomb, while only 14 percent wanted to place it in the hands of the new United Nations. Six weeks later the figures were largely the same: 71 percent of those interviewed preferred maintaining American control, while 17 percent favored supervision by the United Nations. A survey of members of Congress showed even more resistance to a cooperative approach: thirty-nine Republicans and thirty-seven Democrats preferred to hold on to the secret of the bomb; only five Democrats were willing to turn it over to the United Nations. Having won the war, most

Americans, both in and out of the political realm, wanted to retain whatever advantage they possessed as they faced the postwar world.[27]

As he struggled with the issue of secrecy, Truman was clearly sympathetic to the majority stance. Despite the counsel of scientists who predicted that the Russians would be able to build their own bomb in four to five years, he accepted the estimate of General Groves who claimed that the secret was secure for twenty years. While on a fishing trip in October, Truman held a press conference on the porch of his lodge and responded to reporters' questions. Asked whether the United States was prepared to share the secret of the bomb, Truman hedged. He had just underscored the need for some form of international control of atomic energy in his message to Congress the week before, and he wanted to appear accommodating, but he understood the public mood and did not intend to give too much away. The basic scientific knowledge behind the bomb was widely held, he said. "It is only the know-how of putting that knowledge practically to work that is our secret." When pressed to disclose how much information about the bomb he would be willing to communicate, he answered, "Not the know-how of putting it together, let's put it that way." For all his rhetoric, Truman found himself closer to Byrnes and his military advisers than to the scientists and their supporters. And, like the military officials, he perceived no real contradiction between maintaining secrecy and considering collaboration with other powers.[28]

Truman made good on his pledge to Congress to explore the possibility of some form of international oversight. Promising discussions with other nations, he charged Byrnes with handling the American side. Byrnes, in turn, spoke with science adviser Vannevar Bush, who sketched out in November a three-tiered proposal for cooperation with Russia under the auspices of the United Nations. He suggested scientific exchange first, then technical exchange, and finally an agreement to eliminate weapons entirely. An inspection system would protect against surprise attack as the United States shared information in a series of carefully orchestrated stages. At a December conference of foreign ministers in Moscow, the major powers agreed to work through a U.N. commission, and the next month the organization created such a body to deal with the problems of atomic energy.[29]

Despite his eagerness to act, Truman was not entirely sure what course to take. Unwilling to do all the scientists asked, he still had to decide how open he was willing to be, and until he did so, American policy remained unclear. "I got the impression," Oppenheimer later recalled, "that we did not know what international control was or what we could say to the Russians." David Lilienthal noted as well that "those charged with foreign policy—the Secretary of State (Byrnes) and the President—did not have either the facts nor an understanding of what was involved in the atomic energy issue."[30]

Meanwhile, Congress was pressing the President. Members of a new Senate Atomic Energy Committee naively feared that Byrnes was likely to disclose atomic secrets to the Russians in Moscow. "We consider an 'exchange' of scientists and scientific information as sheer appeasement," Senator Arthur Vandenberg of Michigan wrote as he summed up their concerns. They also feared that the contemplated disclosure stages would unfold without adequate safeguards in place. Finally, they wanted a greater role for Congress in working out any arrangement. "I should think he would be God damned glad to consult with Congress before negotiating agreements," Vandenberg said of Truman. "I wouldn't think any one human being would take the responsibility for settling this issue."[31]

Having made a commitment to work through a new U.N. Atomic Energy Commission, top policymakers recognized that they had to define their position on the ticklish question of international control. To that end, Byrnes appointed a committee to frame American policy, with Dean Acheson, second-ranking official in the State Department, serving as chair. Able, articulate, and arrogant, Acheson had helped devise the arrangement whereby Roosevelt had traded obsolete destroyers for British bases in 1940 and had subsequently held a series of government posts that had given him a measure of experience in negotiating with the Russians. Already interested in the atomic issue, he knew how to work with the bureaucracy to accomplish his assigned task. Serving with him on the committee were Vannevar Bush, James Conant, Leslie Groves, and John J. McCloy (one of Stimson's trusted assistants in the War Department). David Lilienthal, then chairman of the Tennessee Valley Authority, headed a board of consultants that included Robert Oppenheimer and a number of corporate executives with scientific expertise.[32]

The committee faced a formidable assignment. It was not enough simply to outlaw the bomb, for other nations would hardly be satisfied with an American pledge not to use the new weapon. An alternative arrangement was necessary, but difficult questions had to be resolved before any pact was possible. If other countries could, in fact, be cajoled into an agreement to forego developing the bomb in return for nuclear information now held by the United States, how could that arrangement be policed? What kinds of inspection might be acceptable, and how intrusive might they be? Lilienthal insisted that his group not "fall into the illusion that there is 'a' solution, one answer that we must seek that will answer everything," yet he remained optimistic that a solution could be found. As he flew around the country to examine various parts of the Manhattan Project, he became increasingly excited about what he learned. "No fairy tale that I read in utter rapture and enchantment as a child, no spy mystery, no 'horror' story, can remotely compare with the scientific recital I listened to for six or seven hours today," he wrote in his journal at the end of January 1946. He was fascinated at "the utter simplicity and yet fantastic complexity of the peering into the laws of nature that is the essence of this utterly bizarre and, literally, incredible business." Intent on finding a solution to the problem of control, he wanted above all "to develop a position, based on facts not now known by our political officers, that will 'work,' and have a good chance of being accepted, especially by Russia."[33]

Oppenheimer provided the formula at the heart of the final report. As in Los Alamos, "the most stimulating and creative mind among us was Robert Oppenheimer's," Acheson later recalled. The physicist, once more at the center of all atomic activity, waited until his colleagues had the necessary background information, then proposed an international agency—an "Atomic Development Authority"—that would maintain a monopoly over the raw materials necessary for atomic reactions. Its advantage was that it would be more than a police authority; it would wield negative powers to stem violations as necessary, but its emphasis would be on its positive responsibilities of promoting peaceful development.[34]

Both the board of consultants and the full committee embraced Oppenheimer's proposal. In the course of revision, despite objections by Lilienthal's group, the committee accepted the idea of a series of stages that would begin with a survey of materials and culminate with the surrender of weapons. It noted that inspection alone would not maintain nuclear peace and acknowledged that too intrusive a system would not work but stressed that some form of inspection was indeed necessary. Finally, the committee endorsed what came to be known as the Acheson–Lilienthal report and sent it to Byrnes. As flexible and accommodating as framers of the plan sought to be, the proposal still left the Soviet Union at a disadvantage, for the United States retained the capacity to produce weapons throughout the entire collaborative process. Further, as Oppenheimer noted at the time, the terms requiring open disclosure of fissionable materials were not always compatible with Russian society, which was more closed and secretive than American society. Still, it was a remarkable step forward, and Acheson's letter of transmittal was correct in underscoring that the group saw the report "not as a final plan, but as a place to begin, a foundation on which to build."[35]

Many of the atomic scientists were pleased, for the report seemed to endorse the approach they had been suggesting since the final days of the Manhattan Project. Edward Teller referred to it as "the first ray of hope that the problem of international control can, actually, be solved." Chemist Harold C. Urey called it "the most statesmanlike pronouncement . . . on the subject since the atomic bomb fell on Hiroshima." William Higinbotham of the Federation of American Scientists was similarly optimistic. "For the first time since the end of the war, we began to feel hopeful," he later wrote. "We clasped the new bible in our hands and went out to ring doorbells."[36]

Not all Americans were encouraged. As the public became aware of the efforts of accommodation, some critics claimed that the United States was not making a good-faith effort to share its secret at all. Journalist I. F. Stone was among those who argued that the plan circumscribed Soviet efforts to explore atomic energy while it preserved America's right to build and test new bombs, and so it was less generous than it seemed. As long as the proposal guaranteed the United States the right to maintain its arsenal and only later hand over the secret, it had little chance of heading off an arms race. The report, Stone concluded, "may turn out to be a prize phony, a slice of atomic pie in the sky." So, too, was poet Robert Frost skeptical. In "U.S. 1946 King's X," he used verse to assail the American approach to atomic energy:

Having invented a new Holocaust,
And been the first with it to win a war,
How they make haste to cry with fingers crossed,
King's X—no fairs to use it any more!

Sharp though their criticism was, it went largely unheard, as the President and his top aides pursued their own chosen course.[37]

Both critics and supporters of the Acheson–Lilienthal report were disturbed by Truman's appointment of Bernard Baruch as ambassador to the recently created U.N. Atomic Energy Commission which would soon deal with the question of control. The wealthy Wall Street financier had long played an important role in government affairs—as director of industrial mobilization in World War I, and as economic consultant in World War II—and relished his reputation as presidential adviser and popular sage. Vain and pompous, frail and hard of hearing at the age of seventy-five, his stature rested on the astute use of public relations and the careful cultivation of those in power. But he knew little about atomic energy and hardly seemed an appropriate choice to present the case for international control. When Lilienthal heard about the appointment, he was "quite sick." In his journal he noted: "We need a man who is young, vigorous, not vain, and whom the Russians would feel isn't out simply to put them in a hole, not really caring about international cooperation. Baruch has none of these qualifications." Oppenheimer later remembered that when he heard about the appointment, he lost all hope. Bush called Baruch "the most unqualified man in the country for the task." For those already critical, the appointment was simply a further indication that the administration had no intention of making a serious effort to work with other nations.[38]

Byrnes had recommended his fellow South Carolinian—a close personal and political friend—in a calculated effort to secure congressional support, and members of Congress who had benefited from his largesse were pleased. Vandenberg indicated that Baruch could skip testifying before the Senate Foreign Relations Committee as long as he promised not to permit disclosure of atomic secrets without adequate safeguards and agreed to submit all pacts to the Senate for approval. Though he accepted his Secretary of State's recommendation, Truman had little affection for Baruch. The day he offered him the post, Truman scrawled on his appointment sheet, "Asked old man Baruch to act as US representative on UNO Atomic Committee. He wants to run the world, the moon and maybe Jupiter—but we'll see." In time, Truman became even more critical, and as opposition to the appointment became more vocal, Byrnes himself acknowledged that the choice was "the worst mistake I have ever made."[39]

The problem was that the egotistical Baruch insisted on redrafting the proposal to his own liking. Irritated at the outset when leaks prompted full publication of the Acheson–Lilienthal report, he had no intention of simply serving as a "messenger boy" if policy had already been set. When he asked who was responsible for shaping the final proposal, Truman responded, "Hell, you are!" The President's assurances that the report was simply a working paper placated him and gave him the opportunity to change it as he chose. Facing reporters who pressed for his view of the Acheson–Lilienthal document as he left a meeting with Truman, Baruch simply smiled, pointed to his hearing aid, and said conveniently, "I can't hear you."[40]

Baruch initially made an effort to learn about the technical issues. He spoke with a number of scientists, Oppenheimer among them, and with Groves, yet he was not really eager to take advice. He said, according to Lilienthal, that "he wasn't much on technical scientific stuff, but he could smell his way through it—and that's the way he did things, smell his way." Baruch decided finally that he did not need the scientists any longer; he knew all he needed to know about the bomb: "It went boom and it killed millions of people and I thought it was an ethical and political problem and I would proceed on that theory."[41]

As members of his staff worked with Acheson, Baruch came to accept the broad outlines of the original report. He endorsed the idea of an international agency, though now it would own only processed uranium and neither the ore itself nor the processing plants. But he made a number of significant changes on the issue of enforcement. He wanted firm penalties for violations, with no Security Council veto possible on questions of punishment. Truman backed Baruch, for he had long felt that the veto impeded adequate enforcement. That decision, which challenged guarantees in the U.N. charter and Russia's intention of using the veto to protect its own interests on this issue, set the United States and the Soviet Union on a collision course.[42]

On June 14, Baruch presented the new American plan at the opening session of the U.N. Atomic Energy Commission. Speaking in the gymnasium of Hunter College in New York, he began on an apocalyptic note: "We are here to make a choice between the quick and the dead." In the next hour, he used extravagant rhetoric to describe "a secret so vast in its potentialities that our minds cower from the terror it creates" and to plead for an end to the peril it brought. Baruch outlined the proposal he had helped refine, with its series of stages leading to dismantling of existing bombs, its system of punishments and controls, and its elimination of the veto by the Security Council on matters of enforcement.[43]

Public reaction was initially favorable. One survey reported that 98.5 percent of press opinion was sympathetic. Military analyst Hanson Baldwin called the plan "thoughtful, imaginative and courageous," while journalist Anne O'Hare McCormick found it "epoch-making . . . as an instrument of world disarmament." Despite their earlier criticism of Baruch, Bush and Lilienthal thought it was a good start, and many of the atomic scientists agreed. But not all response was positive. Higinbotham, like Oppenheimer, was disappointed that so many people accepted Baruch's proposals without understanding all the implications. Journalist Walter Lippmann argued that Baruch's position on the veto was ill-advised, and the Hearst press called the entire plan "imbecilic."[44]

Five days later, Soviet representative Andrei Gromyko responded with a counterproposal. Without referring to the American plan, he suggested a treaty banning production, storage, and use of atomic weapons, destruction of all existing bombs within a three-month period, and a system of penalties for violations. He called as well for the assistance of the United Nations in promoting exchange of scientific information. Finally, addressing the Baruch plan, he stated that the Soviet Union would never agree to eliminate the veto.[45]

The two nations were at an impasse. The American option of international inspection and control, unimpeded by the veto, stood against the Soviet alternative of more flexible national self-control. The Russians, concerned about the continuation of the American nuclear monopoly in the foreseeable future, dug in on the question of the veto and refused to budge. Toward the end of July, a more aggressive Gromyko denounced the Baruch plan and repeated the Soviet proposal. The gap between the two nations grew even larger than before.[46]

Baruch was disturbed. At first he followed Truman's advice to "stand pat," in the hope that the Soviets would come around once they carefully considered the American proposal, but by midsummer he ran out of patience. Aware that an agreement was unlikely, he decided to plunge into the propaganda battle, to persuade others that the Russians were responsible for the failure to reach a settlement. With official support, Baruch decided to ignore the advice of a number of nations counseling delay and to press for a vote in the U.N. commission. On December 30, the resolution passed by a vote of ten to zero, but the abstention of Russia and Poland ended any real hope of acceptance of the American plan. Five days later, Baruch resigned, proud that he had won his personal battle, even if he had lost the larger war.[47]

So ended the American effort at international control and the scientists' dream that collaborative efforts could forestall a deadly arms race. With the standoff in the United Nations, Niels Bohr's dream of a more peaceful world was shattered for good. Baruch's intransigence certainly contributed to the stalemate, but Cold War conflict made any full resolution unlikely. The Soviet Union was jealous of the American lead in atomic affairs and was unwilling to accept any approach that maintained that advantage. The United States, in turn, feared the Soviet Union's conventional troop strength, particularly in eastern Europe, and was unwilling to give up the one weapon that might counter that force. Mutual mistrust precluded both nations from taking the risks necessary to reach an agreement. Under those circumstances, continuing discussions in the United Nations in 1947 and 1948 were moot, and when the U.N. Atomic Energy Commission expired in mid-1948, the prospect of meaningful international control had long since disappeared.[48]

With an international agreement impossible, the United States decided to deal with questions of atomic energy control on its own. The crucial issue in the debate already under way at home was whether military or civilian authority should prevail. Planning for nuclear management had begun in the War Department as early as 1944 and led in the fall of 1945 to the May–Johnson bill, introduced by Representative Andrew J. May of Kentucky and

Senator Edwin C. Johnson of Colorado. It provided for a part-time, nine-member commission which would control all research and production activities associated with atomic energy. The commission, appointed by the President but largely insulated from possibility of removal, could include members of the military. It was responsible for appointing its own administrator, who might also belong to the military. The bill, in short, envisioned a permanent Manhattan Project with continued military control.[49]

Scientists were split on the measure. Many in positions of authority—Bush, Conant, Fermi, Arthur Compton, and Oppenheimer—endorsed the bill, for they were confident that a tolerable relationship with the military could be maintained. But others resented the possible military domination, which they had struggled against so vigorously during the war. They also disliked proposed security restrictions on nuclear research that seemed to them to threaten the free flow of information that science demanded. Finally, they objected to the stress they perceived on military uses of atomic energy and to the message that emphasis sent to the nation and the rest of the world. They agreed with Herbert Anderson, a Metallurgical Laboratory physicist, who declared, "I must confess my confidence in our leaders Oppenheimer, Lawrence, Compton, and Fermi . . . who enjoined us to have faith in them and not influence this legislation, is shaken." Many of them actively lobbied congressmen and members of the administration in an effort to shift opinion against the bill. This time they were more successful.[50]

Without understanding all the implications, Truman initially supported the measure. Characteristically, he sought to act with a decisive stamp. "I want this bill right away," he said. "I need it. We are behind two months on this right now, and a subject so vital can not wait any longer." At a press conference in mid-October, he called the bill "satisfactory . . . substantially in line with the suggestions in the message" he had transmitted to Congress earlier that month.[51]

Truman changed his mind as opposition began to grow. Within the administration, Don K. Price at the Bureau of the Budget and James R. Newman at the Office of War Mobilization and Reconversion argued that the bill deprived the President of the necessary executive control and did not offer an imaginative solution to a revolutionary issue. Budget Director Harold D. Smith similarly told the President that the commission should be wholly under presidential direction. Responding to pressure from both bureaucrats and scientists, the Senate declined to send the bill to the Military Affairs Committee, where it faced a sympathetic hearing, and created instead a Special Committee on Atomic Energy. The hopes of proponents for quick passage evaporated in the face of heated debate.[52]

In response, Truman embraced the idea of civilian rather than military control. At the end of November, he wrote that the May–Johnson bill should be amended to permit civilian supremacy. Several days later, in early December, he called a meeting that included the Secretaries of War and the Navy, at which he first polled the participants and then declared that he believed the proposed Atomic Energy Commission should be under civilian control.[53]

In Congress, Senator Brien McMahon of Connecticut introduced a new bill providing the civilian direction Truman wanted. McMahon, a New Deal Democrat serving his first term in the Senate, became a fierce advocate of a liberal civilian program responsive to the American political system. He was, supporters acknowledged, "one of the sharpest, smoothest apples" on Capitol Hill, someone best not to have against you. Though Arthur Vandenberg helped mobilize support for the measure, at the start, James Newman noted, McMahon "stood alone and fought alone for those essential features of the legislation that were finally upheld."[54]

Acrimonious debate continued until the end of July. Finally, after an amendment creating a Military Liaison Committee to advise a civilian Atomic Energy Commission (AEC), Congress passed the Atomic Energy Act of 1946. The measure established a strict government monopoly over all nuclear materials and facilities, with the government holding all licensing and patent rights. Most important, it underscored the principle of civilian control. A five-member AEC, appointed by and responsible to the President, would direct the entire operation.[55]

To staff the new AEC, Truman appointed David Lilienthal as chairman, and rounded it out with Sumner Pike, a businessman; Lewis Strauss, a financier and admiral in the Naval Reserve; William W. Waymack, a newspaper editor and Federal Reserve branch director; and Robert F. Bacher, a physicist from the Manhattan

Project. All were nonpartisan nominees; indeed, four of the five—
Pike, Strauss, Waymack, and Bacher—were Republicans. Even
so, confirmation proved more troublesome than anticipated. Truman had to fight for Lilienthal's appointment in particular, for the
New Dealer encountered heavy opposition from an increasingly
conservative Congress. In a memorandum he did not release, the
President lashed out at the "peanut politicians" who were causing
him problems once more. But he won that fight, and Lilienthal
and the others were confirmed. Now the nation was ready to
proceed with peacetime nuclear development.[56]

By mid-1946, the United States had resolved the issue of
nuclear control. Faced with the question of what to do with its
terrible new secret, the nation ignored the advice of the scientists
who had produced the bomb. These scientists had entered into
the national dialogue and for a time thought they could carry the
day with their proposal for sharing the secret they had unlocked,
but they found themselves doomed to disappointment, their dreams
of a different world denied. American policymakers—Truman and
his top advisers—reacted to them with the same impatience Winston Churchill had shown when he had spoken with Niels Bohr
during the war. Despite reasoned arguments and emotional pleas,
the scientists failed to persuade those in authority to try a bold
new course, for the nation's leaders saw the world more narrowly
and were unable to take the gamble the scientists advised. Rather
than sharing the secret and collaborating on future development,
the United States elected to proceed with nuclear development on
its own, regardless of the consequences. Agreement with the Soviet
Union might well have been impossible, given the passions of the
Cold War, but the alternative the scientists sought was never tried.
Discouraged, some of them left atomic-energy work altogether;
others resolved to take advantage of the opportunities that lay ahead
and to try to shape nuclear policy that way. Whatever their mis-
givings, they knew that the national and international framework
was now set.

3

Strategy, Weaponry, and the Early Arms Race

The atomic bomb forced a reconsideration of American military strategy. Even while struggling with the question of international control of atomic energy, the United States contemplated the alternative question of how the extraordinary weapon could contribute to national defense. Once international control proved impossible, the nation integrated the bomb into its arsenal, developed new ways to improve its yield, and devised more sophisticated scenarios for its use. Strategic advances that made the bomb the linchpin of American defense accompanied advances in technology, as vastly more powerful hydrogen weapons replaced relatively primitive atomic bombs. The dialogue between scientists and policymakers that had first focused on the issue of nuclear control now broadened to embrace the strategic possibilities of thermonuclear weapons and often unfolded before a public that was alternately pleased and puzzled by the ominous potential of the new bombs.

Scientists remained active participants in the nuclear debate. Many of them—Oppenheimer included—were disappointed at their inability to rein in the force they had unleashed. Some sounded an early warning about the strategic implications of the new bomb. On August 17, 1945, a little more than a week after the destruction of Hiroshima and Nagasaki, Arthur Compton, Enrico Fermi, Ernest Lawrence, and Oppenheimer wrote to Secretary of War Stimson to voice their fears. The development of even more effective weapons "would appear to be a most natural element in any national policy of maintaining our military forces at great strength," they declared. "Nevertheless, we have grave doubts that this further development can contribute essentially or permanently to the prevention of war. We believe that the safety of this nation— as opposed to its ability to inflict damage on an enemy power— cannot be wholly or even primarily in its scientific or technical prowess. It can be based only on making future wars impossible." A number of scientists reacted to what they had done by returning to their laboratories and pursuing nonmilitary projects; more accepted Cold War assumptions and aided the defense effort by helping to develop bigger and better bombs. "Nuclear explosives," physicist Freeman Dyson once observed, "have a glitter more seductive than gold to those who play with them." Even when they had qualms about their efforts, these scientists found themselves fascinated by the technological challenge of their work. Ted Taylor, designer of a massive fission bomb, noted the irony that what he considered his most frightening invention was at the same time the most interesting. Nuclear weapons continued to exert a powerful appeal on scientists after the first secrets of the atom had been unlocked.[1]

Meanwhile, governing officials retained the tight hold over nuclear strategy they had asserted as the war drew toward an end. They relied on strategic analysts rather than on scientists when they wanted advice. These theorists began to reexamine old ideas about the nature of warfare and to frame new assumptions about how the nation might be defended with atomic weapons. They soon understood that the bomb was not simply a larger explosive but a different weapon altogether and shared their growing awareness of new strategic alternatives with the policymakers—at both presidential and staff levels—who had final authority for deciding how the bomb might be used.

These officials debated nuclear possibilities against the backdrop of the Cold War. As they weighed new alternatives, they were never able to ignore the fragile postwar military balance and the polarization that developed between East and West. Many were convinced that the bomb gave them the trump card they needed to prevail. Some, however, were less sanguine about the new strategic prospects. Alexander Sachs, still serving as an occasional adviser, cited the case of sixteenth-century mathematician and inventor John Napier, who created various weapons of war, including a primitive tank and a reflexive burning mirror. On finding that one invention killed an entire flock of sheep some distance away, he destroyed the device to keep mankind from having to confront such lethal power. Though some policymakers hoped to deal similarly with the nuclear secret, more were willing to embrace the new weapon and to rely on it to create the postwar stability they sought. They agreed with Winston Churchill that if the threat was sufficiently frightening, "we shall by a process of sublime irony have reached a stage in this story where safety will be the sturdy child of terror, and survival the twin brother of annihilation."[2]

Some of the nation's top leaders moved quickly to act on the assumption that the new bomb could serve as the lever to force concessions from contentious allies and the cudgel to help keep the peace. James Byrnes, Truman's choice as Secretary of State, was particularly intent on exploring the possibilities of atomic diplomacy. Soon after Roosevelt's death, Truman noted in his *Memoirs*, Byrnes declared that "in his belief the bomb might well put us in a position to dictate our own terms at the end of the war." The next month, Leo Szilard recalled, Byrnes conveyed his "assumption that rattling the bomb might make the Russians more manageable." At a meeting of Allied foreign ministers in London in September 1945, he was determined to use the atomic bomb as a bargaining chip, though he was unclear just how it might be used.[3]

The military, likewise, welcomed the new weapon, though not without an initial period of discontent. Each service sought at first to protect its conventional role from the bomb that had foreclosed other means of ending the war. General Curtis LeMay, for example, called the bomb "the worst thing that ever happened" to the Army Air Forces, but he and others quickly changed their tune. In the fall of 1945, General Carl Spaatz, who had planned the bombing raids on Germany, proposed building the postwar Air Force around the atomic bomb. An Air Force report, drawing on the successful wartime experience with strategic bombing, accepted the atomic explosive as "primarily an offensive weapon for use against large urban and industrial targets." Outlining in January 1946 the role atomic weapons could play, General Dwight D. Eisenhower endorsed a memorandum from Leslie Groves which concluded: "If there are to be atomic weapons in the world, we must have the best, the biggest and the most."[4]

In short, most officials responsible for national defense were soon overcome by the bomb's magnetic appeal. America was at peace for the first time in five years, and the weapon was a visible sign of its authority and might. It was, to civilian and military policymakers alike, a symbol like the ancient Roman fasces that reflected both the inclination and the ability to keep the peace.[5]

Foreign policymakers, however, quickly found that their most optimistic hopes for the bomb were unfounded. During the September 1945 London conference, Byrnes discovered that the new weapon was a less effective master card in diplomatic negotiations than he had hoped. Its real value lay in its understated potential, with American intentions left deliberately vague. But on the third day of the conference, Soviet Foreign Minister Molotov undermined the American effort to maintain the weapon as an ominous background element by asking Byrnes outright if he had "an atomic bomb in his side pocket." Taken aback, Byrnes responded: "You don't know southerners. We carry our artillery in our pocket. If you don't cut out all this stalling and let us get down to work, I'm going to pull an atomic bomb out of my hip pocket and let you have it." The Russian move of drawing attention to the bomb foiled the American effort to use it as a bargaining threat.[6]

Military planners faced a similar problem. Even as they accepted the bomb into the American arsenal, they realized that it might well fail to forestall or check a Soviet attack. Like their diplomatic counterparts, they wondered how it could be used to achieve specific ends, given its apparent limitations for wartime use. Walter Lippmann, well-known political theorist, wrote a series of columns in 1946 outlining the military dilemma. If Soviet troops marched into western Europe, he observed, the United States faced a number of unsatisfactory choices. Nuclear bombs

dropped on western Europe would kill an unacceptably large number of civilians, while bombs dropped on Russia would still leave
the Red Army in the west. For all the hopes it fostered, the new
weapon hardly guaranteed military supremacy.[7]

Another major problem in the first several postwar years was
that the United States was ill-prepared to fight a nuclear war. The
nuclear arsenal was exceedingly thin. The stockpile, which held
only two weapons at the end of 1945, increased to but nine in
mid-1946, to thirteen in mid-1947, and to fifty in mid-1948. The
bombs, all unassembled "Fat Man" implosion weapons, weighed
ten thousand pounds each, and took thirty-nine men two days to
ready for use. A bomb could only be loaded into a bomber by first
moving the weapon into a pit and then using a special hoist to
raise it into the belly of the plane. A quick response to a real or
perceived threat was hardly practical. Nor was there a glut of
fissionable material. Peacetime efforts to promote efficiency and
economy led to the closing of a number of processing plants. The
plutonium production reactors at Hanford, Washington, which
provided the material for implosion weapons, could at any time
have serious problems as a result of sustained operation. In early
1946, one of the three reactors was shut down while the others
were slowed, which caused production of plutonium to fall off
significantly from the wartime rate.[8]

Top policymaking officials were not always sure of the size of
the nuclear arsenal. Truman received no official word about the
stockpile between the fall of 1945 and the spring of 1947, although
Army Chief of Staff Eisenhower evidently briefed him informally
in September 1946 after receiving reports from Leslie Groves. Air
Force and Navy officials were similarly uninformed. As of early
1947, neither the Secretary of the Navy nor the Chief of Naval
Operations knew the size of the stockpile, despite the assumption
of each that the other had been told. With that lack of information,
it was hard to contemplate serious strategic initiatives involving
the atomic bomb.[9]

The first steps in devising a workable nuclear strategy came
as a number of theorists began to reconsider basic assumptions
about war and defense. For a century, strategists had clung to the
dictum of Prussian officer Karl von Clausewitz that "war is a mere
continuation of politics by other means" and had accepted war as
a legitimate instrument of national policy as long as it was fought

for specific objectives. Atomic weapons clouded the picture by minimizing the possibility of recovery for all sides once objectives had been attained.[10]

Bernard Brodie, a 35-year-old associate professor who had just joined the faculty at Yale, was one of the first strategists to consider the impact of the atomic bomb. Brodie had formerly focused on naval warfare. His dissertation, *Seapower in the Machine Age*, had sold well once published, another book of his had become a textbook at the Naval Academy at Annapolis, and a volume on the comeback of the battleship was about done. As soon as he heard about Hiroshima, he understood that the stakes in modern warfare had changed. Telling his wife that all of his past work on the relationship between technology and warfare was obsolete, he turned his full attention to the atomic bomb. A month after the war with Japan ended, he wrote that Hiroshima "heralds a change not merely in the degree of destructiveness of modern war but in its basic character." In the same unpublished essay, he argued: "The atomic bomb is not just another and more destructive weapon to be added to an already long list. It is something which threatens to make the rest of the list relatively unimportant."[11]

Brodie soon elaborated on this theme in *The Absolute Weapon: Atomic Power and World Order*, a book including his and his colleagues' reflections on the bomb. The bomb, he claimed, brought unprecedented devastation, which was all the worse since it could occur anywhere without warning. There was little chance of devising an effective defense against the new weapon, less chance of eliminating it altogether. Instead, it was necessary to learn to live with the bomb. But Brodie was not entirely pessimistic. He hoped that the heretofore unthinkable destructiveness might make the bomb "a powerful inhibition to aggression" if several powers possessed the weapon and so offset one another. His notion of deterrence, rooted in ancient history, revolved around that hope that the bomb would replace the conventional military aim of defeating an enemy with the new goal of preventing an attack altogether. In his classic formulation of the concept of deterrence, which dominated strategic thinking in the decades that followed, he declared: "Thus far the chief purpose of our military establishment has been to win wars. From now on its chief purpose must be to avert them. It can have almost no other useful purpose."

In his view, the United States was caught in the paradox of having to prepare for a war it did not plan to fight.[12]

As Brodie's ideas took hold, the bomb came to play an increasingly important, and a more realistic, part in military planning. Initially, military figures had seen the bomb as simply a more sophisticated conventional weapon and had somewhat naively anticipated using it just as they had used conventional explosives during World War II. A year after Hiroshima, an Air Force plan had proposed dropping fifty atomic bombs on twenty Russian cities—at a time when the arsenal had fewer than a dozen such weapons. In time, the military came to understand the special utility of the bomb, and top service officials began to see how it could fill an unanticipated void. They were concerned about rapid troop demobilization, which cut the ten-million-man wartime Army down to three million by mid-1946, and about economic dislocations—reflected in strikes and rising unemployment rates—which created pressures to cut back military spending. As relations with the Soviet Union deteriorated and the United States committed itself to a policy of containment, the bomb, funded separately from the military budget, became a centerpiece of military thinking almost by default.[13]

Military officials became more aware of the bomb's promise as they watched nuclear tests conducted near the Bikini atoll in the summer of 1946. In the first test, code-named "Able," an atomic weapon was dropped into shallow water more than two miles off target and sank few of the old ships it had been designed to destroy. "Baker," the second test, was more impressive and reassured military leaders of the bomb's value. The Final Report of the Evaluation Board of the Joint Chiefs of Staff the next year noted that "the tremendous destructive (explosive) power of the atomic bomb and the great range of its lethal (radiation and residual radioactivity) effects, combined with the fact that no specific means of defense or of prophylaxis against it exists, or is likely to be devised, make it a weapon capable of decisive importance in war." Used in sufficient numbers, the report suggested, "atomic bombs not only can nullify any nation's military effort, but can demolish its social and economic structures and prevent their reestablishment for long periods of time." Because of the limited fissionable material available, the board recommended that the bomb be used against urban industrial targets and not against naval or troop concentrations. It also endorsed continued development of improved weapons. If the report failed to persuade top officials at this point that the bomb would be the ultimate weapon in a war, it did increase confidence in the weapon's possibilities.[14]

In response, the nation undertook a major, and successful, effort to expand its nuclear arsenal. Over the next several years, Truman approved a series of increases in fissionable material, and technological advances at Los Alamos allowed smaller amounts of plutonium to create atomic explosions. David Lilienthal, first head of the AEC, once noted the pressure to produce ever-increasing numbers of bombs. Whenever he asked for specific requirements, he was simply told to make "more." By mid-1949, weapons had five times or more the explosive power as the initial bomb, and the number of bombs had increased to about two hundred.[15]

At the same time, planners worked to improve delivery systems. The Strategic Air Command (SAC), organized in early 1946, became increasingly important, especially after Curtis LeMay became its head in the fall of 1948. He launched a crash program that by the start of 1949 could deliver 80 percent of the arsenal at any given time. SAC was willing to try anything; as one Air Force officer declared in 1948: "If the Greyhound Bus Company can demonstrate a capability of delivering [atomic] bombs better than any other agency, that company will get the job." Such an alternative was unnecessary. The World War II–vintage B-29, which had carried "Little Boy" and "Fat Man," gave way to the intercontinental, propeller-driven B-36 by 1948 and later to the jet-propelled B-47. In 1955, the B-52 became America's first intercontinental jet bomber.[16]

Strategic planning kept pace with the expansion of the nuclear arsenal. A number of international crises in 1947 and 1948 prompted a greater willingness to consider the use of atomic weapons. The worsening situation in the eastern Mediterranean led the United States to support Greece and Turkey against communist pressure when Britain could no longer provide aid. Meanwhile, the Soviet Union massed troops on the Czechoslovakian border to keep that nation within the communist orbit and refused to allow land access to the western sector of the divided city of Berlin. In those circumstances, the mid-1948 "Intermezzo" plan precluding the use of nuclear weapons was obsolete even before it was completed. In its place, the administration accepted NSC–30 as its "Policy on Atomic Warfare." While the planning document stopped short of declaring publicly that the United States would use nuclear weapons in time of war, the implications were clear. As a State Department analyst observed: "If war of major proportions breaks out, the Military Establishment will have little alternative but to recommend to the Chief Executive that atomic weapons be used, and he will have no alternative but to go along. Thus, in effect, the paper actually decides the issue." Truman declared in September 1948, according to Secretary of Defense James Forrestal, that he prayed he would never have to use the bomb "but that if it became necessary, no one need have a misgiving but what he would do so." Soon after, he endorsed NSC–30. Not all officials were pleased with the unfolding discussion. Lilienthal was one of those who voiced private doubts about American policy and noted in his diary, "Is not the worst fact about modern scientific weapons—notably the atomic bomb—the effect they have upon moral concepts, those patiently built, fragile steps out of the jungle from which man has emerged?" Warfare, he went on, "is no longer conflict within limits imposed by morality, but without limit, without moral containment." Those reservations notwithstanding, the new policy, once adopted, provided the framework for subsequent decisions about how atomic weapons could be used.[17]

Air Force target selection changed as the nuclear arsenal grew. The World War II experience had shown the value of destroying transportation and power facilities, but maps of the Soviet Union were faulty, and the territory was huge and unknown. When specific objectives were hard to identify, planners keyed instead on urban industrial concentrations, as the Joint Chiefs had recommended after the 1946 tests. As even that specificity proved problematical, military officers fell back on a still broader focus. The first SAC plan, prepared in 1948, aimed at population destruction, with industrial destruction a supplementary benefit.[18]

Such plans encountered criticism from both military and civilian officials. One study, coordinated by Air Force lieutenant general Hubert R. Harmon, concluded in the spring of 1948 that bombing seventy Soviet cities would not "*per se*, bring about capitulation, destroy the roots of Communism, or critically weaken the power of Soviet leadership to dominate the people." Such attacks could kill several million civilians and severely complicate life for the survivors but might, at the same time, "produce certain psychological and retaliatory reactions detrimental to the achievement of Allied war objectives." Still, the Harmon report concluded that there were no other military alternatives available.[19]

The next year, Bernard Brodie proved equally critical. Leaving Yale to become a civilian consultant to the Air Force, for which he hoped to play a significant role in shaping atomic strategy, Brodie responded to the request of Air Force Chief of Staff Hoyt Vandenberg to review the target list. Brodie was astonished at the "preatomic thinking" in the services and in the "sheer frivolousness and stupidity" of Air Force targeting. He was disturbed that Air Force officials met only one afternoon a year to determine the principles on which targeting was based and felt that they had really created a "bombing schedule" rather than a carefully crafted war plan. At the same time, he was troubled that planners did not know where power facilities were located or to what extent attacks might damage them. Charging that they were all too willing to make curious compromises without considering their implications, he cited the example of Moscow targeting. Originally, officials had aimed weapons on the Kremlin's spires. On learning that there was an electrical power station a mile away, they selected instead a striking point midway between the two buildings that could have left both buildings undamaged. Finally, Brodie argued that planners had not calculated at all how to destroy Russian war-making ability but "simply expected the Soviet Union 'to collapse' as a result of the bombing campaign" they called the "Sunday punch." Brodie's voice went unheard. Vandenberg looked at his report briefly once while he waited outside in a corridor, then apparently either lost or destroyed it.[20]

Meanwhile, military planners had to face an even more serious challenge as the Soviet Union developed its own atomic bomb. In the charged atmosphere of the Cold War, Americans knew that the Russians would seek to create their own nuclear capability as soon as they could. The only question was how long it would take. Top scientists claimed that with basic scientific knowledge available and with the confidence that a fission explosion was possible, the Soviets could have a bomb in about five years. Vannevar Bush and James Conant subscribed to that timetable; so did Robert Oppenheimer. But other officials argued that it would take the technologically backward Russians far longer. Some people, physicist I. I. Rabi observed, kept pushing the date further and further back: "If you had asked anybody in 1944 or 1945 when would the Russians have it, it would have been 5 years. But every year that went by you kept on saying 5 years." The new Central Intelligence Agency (CIA) claimed in 1947 that because of technical difficulties and scarce raw materials, it was "doubtful that the Russians can produce a bomb before 1953 and almost certain they cannot produce one before 1951." Leslie Groves claimed that the secret was secure for twenty years, and Truman himself was even more confident. In a conversation with Oppenheimer in 1946, his made his view clear:

"When will the Russians be able to build the bomb?" Truman asked.

"I don't know," Oppenheimer answered.

"I know," Truman said.

"When?"

"Never."[21]

He and the rest of the nation were therefore stunned in September 1949 to learn that the Russians had succeeded in their quest. They had not announced their achievement; rather, an Air Force weather reconnaissance plane on a routine mission during the Labor Day weekend had picked up air samples showing higher than expected radioactivity counts. When other samples confirmed the first reading, scientists concluded that the Soviets had conducted a nuclear test. Informed in mid-September of the scientists' judgment, Truman was initially uncertain that the Russians indeed had created a bomb, then chose to delay an announcement for several days to allow a series of other international crises to settle. On September 23, he announced simply: "We have evidence that within recent weeks an atomic explosion occurred in the U.S.S.R."[22]

Suddenly, the United States found that the security in being the world's only atomic power was gone. Worried about being caught unaware, people wondered whether the Soviet test foreshadowed a nuclear attack. At the Capitol, members of the Joint Committee on Atomic Energy struggled to comprehend the implications of the announcement they had just heard. When a thunderclap startled them, someone in the room said, "My God, that must be Number Two!" Physical chemist Harold Urey declared that he was "flattened" by the President's announcement and continued, "There is only one thing worse than one nation having the atomic bomb—that's two nations having it." The editors of the *Bulletin of the Atomic Scientists* moved the minute hand of the "doomsday clock" from seven minutes before midnight to three minutes before midnight to reflect their fear of proliferation.[23]

To Americans, particularly those in policymaking positions, the Russian achievement was a challenge demanding a reply. Cold War competition required a quick response and led the United States to consider how best to maintain its lead in the now-joined nuclear race. Some officials wanted to build more potent atomic bombs. Others proposed developing a different weapon altogether—a far more powerful hydrogen bomb that would simulate the fusion reaction taking place on the surface of the sun. The debate over this new thermonuclear weapon affected all subsequent decisions about the American arsenal and its strategic use in the decades that followed.

The idea of a hydrogen bomb was nothing new. Even as they explored the possibilities of nuclear fission, some Manhattan Project scientists had speculated about an even greater explosion that might be created through nuclear fusion—binding atoms together rather than breaking them apart. Enormous heat was necessary for fusion to take place, and scientists began to reflect that a fission reaction might provide such a source of heat and so trigger an even more powerful bomb. In early 1942, Enrico Fermi had asked, "Now that we have a good prospect of developing an atomic bomb, couldn't such an explosion be used to start something similar to the reactions in the Sun?" He wondered whether deuterons—the nuclei in a heavy form of hydrogen—might undergo fusion at extremely high temperatures and in the process release massive amounts of energy.[24]

Edward Teller was the scientist most passionately interested in creating a hydrogen bomb. The bushy-eyebrowed Hungarian refugee had received his training in physics in Germany and then had gone first to Britain and next to Denmark to work with Niels Bohr when conditions in Germany became inhospitable for Jews. Arriving in the United States in 1935, he carried with him a hatred not simply of Nazism but also of communism, which had prevailed in Hungary in his youth. That hatred fueled his willingness over the next fifty years to do whatever he could to promote American security and defense. Animated, enthusiastic, and interested in everything, Teller loved music—he brought his beloved Steinway piano with him in his wartime work—and was even more willing to dabble with verse. Politics and poetry came together in light-hearted lines he published later in the *Bulletin of the Atomic Scientists* underscoring the need to avoid appeasement in world affairs:

> It seems the Muscovite
> Has quite a healthy, growing appetite.
> We can't be safe; at least we can be right.
> Some bombs may help—perhaps a bomb-proof cellar
> But surely not the Chamberlain umbrellar.

Strong-willed throughout his career, Teller was preoccupied with the idea of a thermonuclear explosion from the moment he first considered it and went to Los Alamos in 1943 determined to work on a hydrogen bomb.[25]

A rift between Oppenheimer and Teller dated back to the middle of World War II. Oppenheimer had asked Teller to work on calculations of the implosion possibilities of the "Fat Man" bomb, but Teller was so absorbed with the possibility of a hydrogen bomb—called the "Super"—that he was of little help. With Oppenheimer's eventual approval, Teller had ended up devoting most of his time to consideration of a fusion weapon. Impatient with the focus on fission, Teller had seemed scornful of the painstaking effort to create an atomic bomb, even after the Manhattan Project achieved its goal. Offered the chance to head the Theoretical Division at the Los Alamos laboratory after the war, he turned down the position with the comment that he was engaged in a more important project. "I'm making an alarm clock," he said. "One that will wake up the world."[26]

Returning to the University of Chicago in early 1946, Teller remained involved with Los Alamos colleagues during regular summer visits. Meanwhile, he continued to proselytize for the hydrogen bomb. In 1947, in the *Bulletin of the Atomic Scientists*, he declared that "it is quite unsound to limit our attention to atomic bombs of the present type." Those weapons had been developed in response to wartime pressure which dictated the course that brought the quickest results. "In a subject as new as atomic power," he went on, "we must be prepared for startling developments. It has been repeatedly stated that future bombs may easily surpass those used in the last war by a factor of a thousand. I share this belief."[27]

Teller's audience became more receptive after news of the Russian bomb. Consulting at the Pentagon in Washington when Truman announced the Soviet test, Teller asked Oppenheimer what he could do. Told simply, "Keep your shirt on," Teller set out on his own to promote a crash program with everyone he saw. While Teller's enthusiasm may have stemmed, in part, from his wish to replace Oppenheimer as the most important scientist in Washington, it was also a result of his rich scientific imagination and of his fervent support of American aims in the Cold War.[28]

Strong support was forthcoming from all sides. Scientists Ernest Lawrence and Luis Alvarez were sympathetic, as were AEC commissioners Lewis L. Strauss and Gordon Dean. Senator Brien McMahon, chairman of the Joint Committee on Atomic Energy, was another champion of the Super. General Omar Bradley, Chairman of the Joint Chiefs of Staff, concluded that if a thermonuclear bomb was possible, the Soviet Union should not be permitted to build it first. Proponents believed that creating such a bomb was morally no different from creating any other new weapon and argued that a decision to forego development would doom the United States to second-class status.[29]

But they faced formidable opposition. Oppenheimer, still the major spokesman for the scientific community, had been troubled by continuing atomic tests in the Pacific and at a new site in Nevada. The detonations excited some scientists and made them jovial and jaunty; they left him "awed and anxious," according to physicist Samuel A. Goudsmit. Oppenheimer felt that a decision to develop the Super could spark an irreversible arms race and that only a decision to renounce such development could offer real hope for world peace. Five of the six scientists on the AEC's General Advisory Committee agreed with Oppenheimer, as did AEC chairman Lilienthal and Commissioners Sumner Pike and Henry Smyth. They argued that a crash program was not the best use of limited resources, that the new bomb was not necessary to maintain national security, and that only an American effort to exercise restraint could persuade others to do likewise.[30]

With the question joined, a major debate among scientists and policymaking officials ensued. It was not a confrontation that pitted one community against the other; rather, scientists faced polarization within their own ranks, and top government leaders had to deal with similar rifts. Nor was it a debate in which the public had a voice. For the most part, the public knew little or nothing about the possibility of a thermonuclear bomb and so played little part in pressuring the other groups. In early November 1949, Senator Edwin Johnson of Colorado, a member of the Joint Committee on Atomic Energy, revealed during a television interview that American scientists were trying to create a far more powerful bomb, and succeeding stories in the *Washington Post* and in the *New York Times* elaborated on his disclosure. While those articles made top officials eager to resolve quickly the matter of whether to proceed with the new weapon, they never gave the public a real role in the deliberations. Rather, the issue unfolded in the AEC, the National Security Council, the Departments of State and Defense, the Joint Committee on Atomic Energy, and the Executive Office.[31]

The first battle came in the General Advisory Committee. At the end of October, the AEC asked the advisory group to meet and determine whether the nation's atomic-weapons program was adequate and whether the development of a hydrogen bomb should be considered. Only Glenn Seaborg, who had discovered plutonium and now conditionally supported the Super, was missing. In an intense two-day meeting, members of the General Advisory Committee (which included some of the nation's top scientists as well as men like James B. Conant, president of Harvard University, and Oliver E. Buckley, president of Bell Laboratories) spoke with military and diplomatic officials and with members of the AEC itself. Finally the committee recommended a number of measures to increase the American nuclear arsenal but stated its unanimous opposition to building a thermonuclear bomb. Although serious theoretical problems had to be overcome, it noted that an aggressive effort might well produce a workable hydrogen weapon within five years. Yet because there seemed to be no limit to the explosive power of the new bomb and no use that could avoid civilian extermination, it declared explicitly: "We all hope that by one means or another, the development of these weapons can be avoided. We are all reluctant to see the United States take the initiative in precipitating this development. We are all agreed that it would be wrong at the present moment to commit ourselves to an all-out effort toward its development." Six of the members of the group,

Oppenheimer among them, appended a statement declaring that "the extreme dangers to mankind inherent in the proposal outweigh any military advantage that could come from this development." Enrico Fermi and I. I. Rabi added their further opinion that because the hydrogen bomb was potentially so powerful, it "becomes a weapon which in practical effect is almost one of genocide."[32]

The AEC itself was split by a three-to-two margin against proceeding with the Super. Lilienthal, one of the opponents, hoped this vote would head off a crash program to develop a hydrogen bomb, for this would only underscore excessive American reliance on weapons of mass destruction, but he knew that the AEC's judgment was not final and understood that the matter would be decided at a higher level. He warned the President in early November that Senator McMahon and the Joint Committee on Atomic Energy would "try to put on a blitz" and was reassured when Truman responded, "I don't blitz easily." Soon after, Lilienthal presented Truman with the positions of the individual commissioners and pressed his own view rejecting development of the new bomb. He declared that existing atomic weapons already provided an adequate deterrent against the Soviets and argued that an effort "to produce something capable of almost *unlimited* destruction" would alienate international opinion and undermine hope for world peace.[33]

Truman responded by appointing a special committee of the National Security Council to help resolve the growing debate and advise on the development of a hydrogen bomb. Consisting of Lilienthal, Secretary of State Acheson, and Secretary of Defense Louis Johnson, the committee reflected the polarization in top circles. Johnson was a strong proponent of the new weapon. Acheson was ambivalent, feeling that a commitment to international control had to be balanced against the needs of national security. Lilienthal remained opposed.[34]

As the various groups deliberated, supporters of the new weapon became more vocal. Massachusetts Institute of Technology president Karl Compton wrote to Truman arguing that the Soviet Union could not be permitted to gain the lead in the nuclear race. McMahon sent the President a rebuttal to the General Advisory Committee report, in which he claimed that a large explosive was necessary to compensate for possible targeting inaccuracy and denied that there was any moral difference between a big bomb and several small ones. "If we let Russia get the super first," he declared, "catastrophe becomes all but certain—whereas, if we get it first, there exists a chance of saving ourselves first." AEC commissioner Strauss agreed that development of a thermonuclear weapon was necessary: "Its unilateral renunciation by the United States could very easily result in its unilateral possession by the Soviet Government." And General Bradley, speaking for the Joint Chiefs, wrote to the Secretary of Defense that it was "intolerable" to allow the Russians to get the bomb first.[35]

Truman appeared sympathetic to the proponents' position. Even before his special committee reported back, Acheson observed, he told the executive secretary of the National Security Council that the positive recommendation of the Joint Chiefs of Staff "made a lot of sense." Although he had listened to members of the General Advisory Committee, they had not persuaded him, and now he saw no alternative but to move ahead.[36]

On January 31, 1950, the President met for seven minutes with the special committee. Its recommendation—a compromise between a crash program and rejection of the Super—was that the President approve a program to determine the technical feasibility of producing a hydrogen bomb. Truman declared that he did not feel that such a bomb would ever be used but that because of the way the Russians were acting, he believed such a program was necessary. When Lilienthal registered his dissent, Truman cut him off. It was, the AEC chairman noted later, like saying " 'No' to a steamroller." Truman only wanted to know, "Can the Russians do it?" When told that they could, his response was, "We have no choice. We'll go ahead." In his public announcement, an effort to play down the issue, he simply said: "I have directed the Atomic Energy Commission to continue its work on all forms of atomic weapons, including the so-called hydrogen or superbomb." Newspapers were not fooled by the calculated effort to minimize the importance of the new policy. The next morning the New York Times ran a bold headline declaring: TRUMAN ORDERS HYDROGEN BOMB BUILT.[37]

While Truman had finally acted with the decisiveness he favored in plunging ahead, his choice was an obvious one. Like the decision to use the atomic bomb, this move was governed by the momentum of events and followed logically from steps already taken. Listening to those scientists whose perceptions were closest to his own, he followed his first instincts. When he considered his alternatives against his perception of America's Cold War needs, "there actually was no decision to make on the H-bomb," he told adviser Eben Ayers a week after the meeting with the special committee. "He went on to say that we had got to do it— make the bomb—though none want to use it," Ayers noted. "But, he said, we have got to have it only for bargaining purposes with the Russians." Once again Truman stood back, let committees air the issue, then gave quick approval to those who shared his view. And once again he allowed the decision-making process to unfold in secret.[38]

Though most Americans, caught up in the Cold War consensus, supported Truman's decision after it became public, not everyone was pleased. Soon after the presidential announcement, Albert Einstein declared that if the hydrogen bomb proved successful, "radioactive poisoning of the atmosphere and hence annihilation of any life on earth has been brought within the range of technical possibilities. . . . In the end there beckons more and more clearly general annihilation." A group of prominent physicists, including Hans Bethe, charged that "this bomb is no longer a weapon of war but a means of extermination of whole populations." Oppenheimer, more circumspect after having been overruled, observed in a radio interview that the hydrogen-bomb decision involved issues that "touch the very basis of our morality" but complained only about the secrecy surrounding the process.[39]

Truman was unconcerned with the criticism. As he signed the prepared statement authorizing the development program, he remarked: "I remember when I made the decision on Greece. Everybody on the National Security Council predicted the world would come to an end if we went ahead. But we did go ahead and the world didn't come to an end. I think the same thing will happen here." Though Truman may have downplayed the catastrophic possibilities, he failed to acknowledge the enormous impact his decision did, in fact, have on the postwar world. His determination to proceed with the development of a hydrogen bomb made it all but certain that the Soviets would embark upon a similar program. By deciding to press on to the next plateau, Truman confirmed America's growing reliance on nuclear weapons when determining strategy. He also moved the nation, and the world, one step further in an escalating arms race with the mistaken assumption that it was possible for the United States to stay in front.[40]

Once the decision was made, development of the hydrogen bomb began. Although Teller and his supporters believed that they had a better than even chance of success, they soon found that their early calculations were wrong. Initially, scientists considered several methods, the most promising of which was the "classical Super," first proposed by Teller in 1942. They envisioned using a cylinder of liquid deuterium and generating fusion reactions between deuterium nuclei with tremendous heat from an exploding fission bomb. The problem was that the temperatures necessary were far higher than an atomic bomb could provide. They then speculated that tritium, another hydrogen isotope, might lead to a fusion reaction at a lower temperature. Several months after Truman's decision, mathematician Stanislaw Ulam calculated that such large amounts of the rare tritium were required that the method was unworkable and suggested that Teller's earlier estimates were mistaken. On hearing the news, Ulam noted, Teller went "pale with fury." No one, Hans Bethe later wrote, blamed him for making mistakes, especially at a time when adequate computers were not available. "But he was blamed at Los Alamos for leading the Laboratory, and indeed the whole country, into an adventurous program on the basis of calculations that he must have known to have been very incomplete."[41]

By the end of 1950, Teller was desperate to find a solution, but none of his ideas showed promise. Then, while working on fission weapons in early 1951, Ulam began to consider compressing deuterium by directing the shock wave of an atomic bomb. When Teller learned of the idea, he proposed focusing radiation, rather than the shock wave, to cause the compression of deuterium. This was the breakthrough he had been seeking. The concept, which Bethe believed was "about as surprising as the discovery of fission had been to physicists in 1939," made further progress possible and turned the project around. As Oppenheimer later noted, "It is my judgment in these things that when you see something that is technically sweet you go ahead and do it and you argue about what to do about it only after you have had your technical success."[42]

The first test of a thermonuclear device—though not yet a

workable bomb—occurred at Eniwetok in the Marshall Islands in the Pacific on November 1, 1952. The Mike shot, part of the Operation Ivy series, was overwhelmingly successful. With a yield of ten megatons (ten million tons) of TNT, it was a thousand times more potent than the first atomic bombs. It left a crater in the seabed a mile long and 175 feet deep and eliminated the island of Eleugelab. In Berkeley with Ernest Lawrence and Luis Alvarez, watching instruments that recorded the impact, Teller apologized for the tremendous destructive force. The bomb's creators, he explained, had worried that if the test failed, Washington might cut off funding for the program, and so they had included every possible design to make the device stronger. "Everything worked," Alvarez observed.[43]

A thermonuclear arms race was soon under way. Fallout debris from the Mike test provided the Russians with evidence of a compressed-material reaction having taken place, and they accelerated their efforts to produce a hydrogen bomb. In mid-August 1953, the Soviets successfully ignited a small amount of lithium deuteride—a different form of thermonuclear fuel—with a large amount of fissionable material. Though they had not yet developed a thermonuclear bomb, they had taken the necessary first steps. The following March, in the Bravo test at the Bikini atoll, the United States exploded a fifteen megaton bomb, also made of lithium deuteride, which was easily adaptable for aircraft delivery. The Soviets followed suit in November 1955 with a hydrogen bomb, dropped from a plane, that was similar to, but smaller than, the one the United States had tested twenty months before.[44]

The stakes rose as the world entered the thermonuclear age. The superpowers took a quantum leap forward as their new hydrogen weapons made the first fission bombs pale by comparison. Now fears of cataclysmic destruction became even more pronounced than in the months after Hiroshima. In 1953, the editors of the *Bulletin of the Atomic Scientists* moved the minute hand of the signature clock even further forward, to two minutes before midnight. In 1954, Val Peterson, head of America's civil-defense effort, notified Lewis Strauss, now chairman of the AEC, of "a growing public feeling of hopelessness or fatalism predicated on a belief that hydrogen weapons are so horrible that there is no defense against them." That same year, British philosopher Bertrand Russell declared in the *Bulletin of the Atomic Scientists* that human beings had never before "been faced with so great a danger as that which they have brought upon themselves by a combination of unrivaled skill and unrivaled folly." Russell feared for man's very survival. "As geological and biological time goes," he concluded, "he has as yet had a very short run. It would seem a pity that he should exterminate himself while still on the very threshold of his possibilities." His countryman Winston Churchill pointed out the newest perils best of all in 1955. "There is an immense gulf between the atomic and the hydrogen bomb," he declared. "The atomic bomb, with all its terrors, did not carry us outside the scope of human control or manageable events in thought or action, in peace or war." With the advent of the world's newest weapon, "the entire foundation of human affairs was revolutionised, and mankind placed in a situation both measureless and laden with doom."[45]

The scientists who struggled to stifle development of the hydrogen bomb before it was too late saw their worst fears confirmed. Like Niels Bohr during World War II, they worried about what might happen without a successful effort to use human reason to temper the consequences of human expertise. Once again, Oppenheimer was a spokesman for his colleagues as he articulated their disappointment in every new step taken in an escalating arms race that left the United States and the Soviet Union like "two scorpions in a bottle, each capable of killing the other, but only at the risk of his own life."[46]

As the United States created ever more destructive weapons, critics could only ponder the road not taken. Then and later, they understood that the new bomb was not necessary to defend American national security; an ample arsenal of fission bombs provided more than enough protection. At the very least, alternatives might have been explored before it was too late. Even after development was under way, there were chances to pause. In October 1952, just before the Mike shot, Vannevar Bush urged the United States to forego testing the new weapon if the Soviet Union would do the same. He feared—accurately, it turned out—that the Russians would learn about the American program from the fallout and felt strongly, he later recalled, that the test "ended the possibility of the only type of agreement that I thought was possible with Russia at that time, an agreement to make no more tests." First an arms control panel, chaired by Oppenheimer, vetoed his proposal, then

Secretary of State Acheson did the same when Bush persisted. No one was willing to let the Russians get ahead. And so at this juncture, as at previous ones, the United States—and the Soviet Union—pushed on to the next level of nuclear development without looking back.[47]

The polarization so evident in the debate over developing a hydrogen bomb continued in the 1950s. The trials of Alger Hiss, who was accused of espionage and ultimately convicted of perjury, made countless Americans anxious about spying that assisted the Soviets, even as others claimed that there was nothing to fear. The arrest, and eventual execution, of Julius and Ethel Rosenberg for passing atomic secrets to the Russians similarly highlighted a perceived threat to national security and caused the same kind of public debate. That polarization came to a head in the hearing over whether or not Robert Oppenheimer was a security risk. Oppenheimer's left-wing associations had long proved disturbing to security-conscious officials, and his clearance to work on the Manhattan Project had come through only at the insistence of General Groves. In the more feverish atmosphere of the Cold War, past ties continued to plague him, and his opposition to the Super provided adversaries with the ammunition they needed to claim that he constituted an unacceptable threat. In the process, other scientists who had similar doubts found that they, too, were no longer welcome in decision-making circles.

The attack on Oppenheimer began in November 1953 with a letter from William L. Borden, former executive director of the Joint Committee on Atomic Energy, to FBI head J. Edgar Hoover. Borden charged "that more probably than not J. Robert Oppenheimer is an agent of the Soviet Union." When President Dwight Eisenhower learned of the accusation, he immediately suspended Oppenheimer's access to all classified information until the issue was settled, and so started a seven-month ordeal during which all parts of the scientist's life were examined. Oppenheimer's denial of the charges led to an exhaustive AEC investigation and a four-week hearing in the spring of 1954 that left him mangled and raw.[48]

Lewis Strauss played the dominant role in managing the inquiry. Strauss was a strong-willed, self-made man, who had begun as a shoe salesman, had served as private secretary to Herbert Hoover, and had become rich as an investment banker before he was called to military service from the Naval Reserve during World War II and became a rear admiral. Named to the AEC by Truman and to its chairmanship by Eisenhower, he often found himself in the midst of controversy and responded when provoked, according to one observer, with an expression that varied "between childish indignation and pouting martyrdom." A committed Cold Warrior and Teller's close friend, Strauss found it difficult to tolerate dissent, particularly on questions surrounding nuclear arms.[49]

Strauss was determined to bring Oppenheimer down. The physicist had once publicly embarrassed the commissioner at an AEC hearing, and Strauss did not forget such affronts. Nor could he countenance what he took to be Oppenheimer's arrogance in hesitating to support the strongest possible program of American defense. To achieve his goal, Strauss had the FBI follow Oppenheimer, tap his phone, and monitor conversations with his attorney. Without acknowledging the source of his material, he provided guidance to the counsel for the AEC and during cross-examination made sure that Oppenheimer faced prickly issues without access to documents that could have aided his case.[50]

The hearing questioned Oppenheimer's communist associations and past improprieties. But the real focus was his opposition to the hydrogen bomb and America's nuclear strategy and the character defects his opposition ostensibly revealed. Though some scientists, like I. I. Rabi and Lee A. DuBridge, spoke on Oppenheimer's behalf, the testimony of longtime nemesis Teller had the greatest impact. Asked whether he considered Oppenheimer a security risk, he responded that on numerous occasions Oppenheimer acted "in a way which for me was exceedingly hard to understand" and said that he "would feel personally more secure if public matters would rest in other hands." Though he did not challenge Oppenheimer's intentions or his ability to keep secrets, he did distrust his decisions, and he declared: "If it is a question of wisdom and judgment, as demonstrated by actions since 1945, then I would say one would be wiser not to grant clearance."[51]

The three-man hearing panel found Oppenheimer at fault. Though it saw "no evidence of disloyalty" and "much responsible and positive evidence of . . . loyalty and love of country," it still concluded that Oppenheimer's disregard for security and his decisions concerning the hydrogen bomb were reasons enough for denial of clearance. Next, the general manager of the AEC upheld the panel's ruling, and finally the commission itself concluded that "concern for the defense and security of the United States requires that Dr. Oppenheimer's clearance should not be reinstated."[52]

The judgment shattered Oppenheimer. He continued to direct the Institute for Advanced Study but had lost his reputation and what remained of his ability to influence American policy in the nuclear age. Though he still enjoyed the respect of scientific colleagues, his standing before the rest of the country was damaged. His friends were furious with Teller for his testimony and refused to speak to him. Some scientists declined to cooperate further with the military. Others sensed the need to be more circumspect in working for the government, particularly on atomic research. They felt, too, a change in their own status, for there was a measure of truth in William Borden's contention that the verdict was "the turning point in converting scientists back to human beings."[53]

The development of the hydrogen bomb served as a more general turning point in atomic affairs. Eisenhower won election several days after the Mike test and had thermonuclear weapons available before the end of his first term. Embracing his predecessor's commitment to nuclear arms, he made these weapons the foundation of American strategic defense.

Eisenhower was well-equipped to deal with nuclear issues. The consummate military leader, he had persided brilliantly over the Allied forces in Europe during World War II. As Army Chief of Staff from 1946 to 1948 and later as Supreme Commander of North Atlantic Treaty Organization forces, he had to consider how the new atomic bomb might be used. Recognizing the weapon's role in military affairs, he never lost his sense of the need to proceed with caution. His reputation as a war hero easily won him election as President in 1952, and his natural, homey manner made him widely popular throughout his two terms. Eisenhower seemed to embody basic American values. Dedicated to preserving the peace, he was equally committed to protecting American security. At the same time, he sought a balanced budget, less intrusive government, and the continuation of the prosperity of the postwar years. Those priorities helped shape his policy as he turned to defense. He drew as well on the views of Secretary of State John Foster Dulles, a devout Presbyterian who hated atheistic communism and was dedicated to doing whatever was necessary to contain Soviet influence, indeed, to roll it back entirely if he could. Together, Eisenhower and Dulles defined a new framework for defense in the atomic age.

In late October 1953, Eisenhower approved a statement of America's national security policy. The document he authorized included a key sentence: "In the event of hostilities, the United States will consider nuclear weapons to be as available for use as other munitions." The nation may have made that implicit assumption ever since the Trinity test; it now became an explicit guideline for policy. The "New Look," with its reliance on nuclear arms, would preclude huge expenditures on conventional forces and so keep the budget in line. In his State of the Union message the following January, the President declared that the United States would deter aggression by maintaining "a massive capability to strike back." Several days later, in a speech to the Council on Foreign Relations, Dulles outlined even more specifically the doctrine that came to be known as "Massive Retaliation." The United States, he announced, would "depend primarily upon a great capacity to retaliate, instantly, by means and at places of our choosing," to counter aggression anywhere in the world.[54]

To make good on its pledge, the United States built more and more bombs. The arsenal, which included a thousand atomic warheads when Ike assumed the presidency, grew to contain eighteen thousand warheads by the time he left office. The advent of new delivery systems made the bombs even more deadly. In 1957, the Soviet Union shocked the United States, first by flight-testing an intercontinental ballistic missile and then by launching *Sputnik I*, a small satellite that went into orbit in space. The Russian achievement made Americans feel vulnerable to attack, for the rocket that lifted the satellite might well carry a hydrogen bomb. Bombers took ten hours to reach enemy targets; missiles could reach their mark in but thirty minutes. Responding to the Russian challenge, the United States sped up its own missile program with the first results evident by the end of the decade. The stakes in the arms race continued to rise.[55]

Despite the ominous sound of "Massive Retaliation," the Eisenhower administration remained vague about how the atomic arsenal might be used and proceeded with restraint. The United States threatened the Chinese with nuclear weapons if they did not agree to end the Korean war, but the President refused to box himself into a position in which he had no further choice, and the weapons never became necessary. Dulles, according to some accounts, offered the French military atomic bombs in the campaign against the Vietnamese, yet in this conflict, too, the weapons remained unused. Throughout his presidency, Eisenhower hoped to control the arms race and insisted on sufficiency rather than needless superiority in making preparations for attack.[56]

All the while, planning for widespread use of nuclear weapons continued. A Strategic Air Command document in 1954 envisioned a tightly coordinated operation against the Soviet Union that could, according to a Navy observer, leave that nation "a smoking, radiating ruin at the end of two hours." Before Eisenhower left office, military officials from the various services produced the first Single Integrated Operational Plan (SIOP) for fighting a nuclear war. That design provided for an "optimum mix" of military, industrial, and government targets for destruction in a single, coordinated attack.[57]

As nuclear planning unfolded in the 1950s, Bernard Brodie remained critical of the American approach. A major attack might well invite a similar Soviet response. Alternatively, if the Soviets felt that the United States was too sane to use such firepower, then the effect of the deterrent was lost. Working at the RAND Corporation—an Air Force–sponsored research organization—Brodie proposed instead that in the event of a war in western Europe, the United States should fire just a few nuclear weapons on invading Soviet forces and none on Russian cities. American leaders could then threaten further attacks on urban areas if the Soviets failed to stop.[58]

Neither military planners nor civilian officials took Brodie's ideas seriously, but other theorists were intrigued. In 1957, Harvard government professor Henry Kissinger published *Nuclear Weapons and Foreign Policy*, in which he claimed that "we cannot base all our plans on the assumption that war, if it comes, will be inevitably all-out." Seeking to define a new relationship between force and diplomacy, Kissinger suggested that limited nuclear war was a legitimate alternative to a massive atomic holocaust. Three years later, RAND strategist Herman Kahn argued much the same thing in *On Thermonuclear War*. In his meandering book, he contended that "for at least the next decade or so, any picture of total world annihilation appears to be wrong, irrespective of the military course of events," and suggested that a limited nuclear war could be fought and won. Rather than shrink from such assessment, policymakers needed to contemplate how each stage of a thermonuclear conflict could be negotiated and survived. Both Kissinger and Kahn were attacked for their work. One critic charged that Kissinger seemed "lyrical" about nuclear war; another termed Kahn's book "deplorable" and called it a "hoax in bad taste," a "moral tract on mass murder: how to plan it, how to commit it, how to get away with it, how to justify it." Despite such attacks, the views voiced carried strategic development still one step further and provided a model for subsequent administrations to consider.[59]

In the fifteen years following World War II, the nuclear bomb assumed a preeminent place in American military plans. Its awesome power, whether sparked by fission or fusion, made it more destructive than any weapon yet known, and gave it a central place in the ever-expanding arsenal. Political judgments limiting spending on conventional arms in both the Truman and Eisenhower years secured its special status and ensured that strategy revolved around the nuclear bomb. But even as planners made clinical assessments of yield and plotted which targets to destroy, fears about increasingly destructive possibilities became more loudly expressed than before. Once again, in their effort to head off the hydrogen bomb, scientists were in the forefront of the attack, and once again, they found other officials unwilling to follow their lead. Still, as development proceeded, their fears could not be denied. As his final term drew to an end, Eisenhower was frightened at the enormous number of targets and at the anticipated overkill that were part of the integrated plan he was asked to approve. After listening to his science adviser, George B. Kistiakowsky, evaluate the proposal, Ike confided that the presentation "frighten[ed] the devil out of me." Other Americans, not cognizant of the most sophisticated military plans, were, nonetheless, anxious about the widespread and highly visible testing that created more awesome weapons and threatened to poison the atmosphere. A public dialogue, focused first on the question of fallout, became an integral part of the thermonuclear age, even as strategic planning proceeded with a life of its own.[60]

4

Fear of Fallout

Fallout became a major issue in the 1950s. Public attention, riveted
first on the possibilities of cataclysmic destruction, shifted toward
less visible but equally deadly nuclear perils as the United States
and the Soviet Union tested increasingly powerful bombs. Long
aware of the dangers of radiation, even as they debated the limits
humans could tolerate, Americans in the first decades of the twentieth century had searched for ways to protect themselves while
still enjoying the benefits of new elements. With the advent of
nuclear weapons, they found that, in spite of all efforts, radioactive
debris from atomic tests was invading their bodies and bones and
threatening their lives. Radiation from fallout, physicist Ralph
Lapp observed, "cannot be felt and possesses all the terror of the
unknown. It is something which evokes revulsion and helpless-
ness—like a bubonic plague." In the first years after Hiroshima,
most Americans were entranced by the promise of atomic energy
and cheered by the apparent security provided by the bomb. In
the 1950s, even as they continued to dream of the potential of
commercial nuclear power, they wondered whether anyone, anywhere in the world, could withstand the invisible consequences
of a nuclear attack. Had they created a Frankenstein's monster

that would kill them in silent, insidious ways? Were they fiddling with the future of the human race by developing bigger and better bombs? Those questions dominated public debate in the second decade of the atomic age. Earlier decisions—dealing with development of fission and fusion weapons—had been made beyond the pale of public view. In the same way, scientists and policymakers had dominated discussion about the possibility of international control. The fallout issue unfolded more openly, as commentators in various fields speculated about the hazards of radioactive bombs. Top officials—and scientists—who had formerly spoken largely among themselves, now had to confront the criticisms that became part of popular culture and helped define the framework for national debate.[1]

The discovery of radiation dated back to the late nineteenth century. In 1895, German physicist Wilhelm Konrad Roentgen had found that a barium-coated screen brightened when he passed an electric current through a glass bulb with most of the air removed. He realized that the tube was producing some kind of invisible radiation, which went through such substances as paper, wood, and even the human body that normally block light. Fascinated, he called the new rays X-rays, with the "X" standing for "mystery" or "unknown." Other scientists quickly began to work with X-rays. American inventor Thomas Alva Edison failed in a quest to take X-rays of the human brain but developed a fluoroscope to give an instantaneous X-ray image. Meanwhile, French scientist Henri Becquerel wondered whether luminescent materials— which gave off light when exposed to sunlight—might also emit X-rays. Working with uranium, he found that the element left an image on a fully wrapped photographic plate. Chemist Marie Curie took Becquerel's discovery a step further. In 1898, she noted that pitchblende, an ore containing uranium, gave off more radiation than could be accounted for by the content of the uranium alone. Referring to spontaneously emitted radiation, she coined the term "radioactivity." That same year, she and her husband, Pierre, identified a new radioactive element—radium—in the pitchblende and then struggled to isolate enough of it to make the measurements necessary for official recognition of the finding.[2]

The American public was fascinated by the discoveries. The *Journal of the American Medical Association* observed in 1896 that European surgeons "believe that the Roentgen photograph is destined to render inestimable services to surgery," and E. P. Davis, editor of the *American Journal of Medical Sciences*, declared at about the same time that X-rays "might prove useful in the diagnosis of pregnancy." Some physicians used them to remove undesired body hair. At the Electric Light Association Exposition in New York in May 1896, Edison let people place hands, legs, and head in front of the rays and watch themselves on the fluorescent screen. The magazine *Photography* carried the poem "X-actly So!" which concluded:

> I'm full of daze
> Shock and amaze;
> For nowadays,
> I hear they'll gaze
> Thru' cloak and gown—and even stays,
> These naughty, naughty Roentgen Rays.

Some years later, researchers at the University of Pennsylvania tried to use X-rays to photograph the human soul.[3]

Radium quickly became as popular as Roentgen's X-rays. Sir William Ramsay, an expert in the new field dealing with radioactivity, suggested that with the identification of radium, the *"philosopher's stone* will have been discovered, and it is not beyond the bounds of possibility that it may lead to that other goal of the philosophers of the Dark Age—the *elixir vitae*." Dancers performed radium dances, their veils dipped in fluorescent salts including radium. Radium roulette in America featured a wheel covered with a radium solution that made it glow in the dark. Dr. W. J. Morton, who had successfully promoted X-rays earlier, now introduced "Liquid Sunshine"—a drink to be taken internally that would become fluorescent when radium was placed near the skin. The assumption was that the glowing internal rays would bathe internal organs and heal any disease. In 1904, at the annual dinner of the Technology Club—the Massachusetts Institute of Technology alumni association in New York City—Morton mounted a "Liquid Sunshine Dinner." At the end of the affair, guests downed their drink with enthusiasm. None "of the self-sacrificing scientists who drank the liquid became transparent afterwards," the *New York Times* reported, "although all were assured that, as a matter of fact, their interiors were thoroughly illuminated."[4]

Even as inflated hopes unfolded, researchers became aware of radioactivity's darker side. As early as 1896, X-ray therapists began to observe serious side effects that accompanied their treatments. In Vienna, Dr. Leopold Freund used X-rays two hours a day for sixteen days to remove a hairy mole from the back of a five-year-old girl. After twelve days, her back became badly inflamed and proved slow to heal. That same year, Clarence Dally, a glassblower who worked for Thomas Edison, started to suffer from skin ulcers following his efforts to produce X-rays inside a calcium tungstate-coated tube. He continued such work for the next two years, eventually trying skin grafts to repair his damaged hands. Nothing helped as the X-ray burns turned into cancer, led to amputations of injured limbs, and finally killed him in 1904.[5]

Publicity accompanied the problems with radiation. In July 1897, the *New York Times* warned of "Roentgen Ray Dangers" in an editorial noting the numerous cases of severe burns. Scientific journals carried similar reports. In one of them, slightly more than a decade later, Dr. John Hall Edwards recorded the intense pain he endured from such burns. "Drugs have so far failed to give me the slightest relief," he wrote. "The pain is of a neuralgic character, it never ceases and is from time to time intensified by sudden stabs and jumps of such severity as to make one cry out."[6]

These victims were suffering from the ionizing effects of X-rays. The high-energy rays knocked electrons out of atoms as they collided and left the atoms with an electrical charge. These ionized atoms proved unstable, combined with other atoms in different ways, and sometimes caused illness or death. Caution became increasingly important as the consequences of careless exposure became better known.[7]

Radium posed similar dangers, vividly dramatized in the third decade of the twentieth century by the experience of women who painted luminous dials on wristwatches. Many of them worked for the United States Radium Corporation, founded in 1916, which produced watches as well as occasional specialty items, like crucifixes that glowed in the dark. Young women, some no more than twelve, did most of the painting. Using a radium-based oil paint, they licked their brushes to a fine point to do the necessary detail work and remove excess paint. Sometimes, for fun, they painted their nails or teeth to make them glow. Each time they tipped their brushes, they ingested minuscule amounts of radium. In the early

1920s, as demand for radium-lit items increased, a number of the young painters died, and many more suffered from serious jaw problems. In 1924, a New York dentist treating one of the women realized that radioactivity was involved in his patient's illness. Subsequent studies revealed that the women often had radon gas on their breath, which foreshadowed serious bone and blood diseases. The women—and chemists and others who worked with the radium—were experiencing the devastating effects of alpha and gamma emissions. Radiation, in both particle and ray form, penetrated their bodies, bombarded organs and skeletal structures, and frequently produced cancers culminating in death.[8]

Professional organizations on both sides of the Atlantic had long called for care in dealing with increasingly evident radiation hazards. In 1913, the German Roentgen Society issued a set of guidelines to protect X-ray workers from exposure, and the British Roentgen Society followed suit two years later. After World War I, British physicians and radiologists established a radiation-protection committee which handed down recommendations for further safeguards from X-ray and radium hazards. As health problems became more visible, the Second International Congress of Radiology organized the International X-Ray and Radium Protection Committee in 1928, which was followed the next year by an American counterpart—the Advisory Committee on X-Ray and Radium Protection. In 1932, the American Medical Association denounced the use of X-rays to remove body hair and, three years later, withdrew approval of internal applications of radium. Over the next several years, the advisory committee prepared further guidelines for X-ray and radium use.[9]

Meanwhile, scientists had explored the question of how much radiation human beings could tolerate. Physicist Arthur Mutscheller, born and trained in Germany but working for an American manufacturer of X-ray equipment, measured doses received by physicians and technicians at several laboratories who had suffered no apparent adverse effects. Relying on his own formula, he produced a number limit that he presented to the annual meeting of the American Roentgen Ray Society in 1924. He worked with an erythema dose—the amount of radiation necessary to cause a reddening of the skin. The next year Rolf Sievert, a Swedish scientist, arrived independently at an equivalent figure, and two British physicists reached similar conclusions several years later. In 1933,

Lauriston Taylor, an American physicist who chaired the Advisory Committee on X-ray and Radium Protection, recommended that his group adopt Mutscheller's figure, and in 1934, both the American advisory committee and the International X-Ray and Radium Protection Committee set tolerance doses, derived from Mutscheller's work. Though the original studies were hardly exhaustive or authoritative, they led directly to standards, expressed in roentgens or units of exposure, that appeared crisp and precise.[10]

The Manhattan Project had made the issue of radiation safety more pressing. Nuclear fission created radioactive isotopes and elements not found in nature and compounded the task of providing protection. Radiation lingered after the Trinity test and the detonations in Japan. Accidents, during and after the effort to create a bomb, highlighted the problem. In 1944, a small amount of plutonium exploded in a chemist's face, and he swallowed some of the substance. When no one was able to measure how much plutonium had been ingested or what the effects might be, scientists demanded immediate research efforts to determine how to detect and gauge the amount of plutonium the scientists on the project harbored. Several other accidents that produced critical reactions were even more ominous. One in early 1945 exposed four workers to gamma and neutron radiation, though all survived. Another, just after the successful detonations at Hiroshima and Nagasaki, led to the death of Harry K. Daghlian a month after he inadvertently achieved criticality in an assembly he was working on alone. A third had a similar deadly effect on physicist Louis Slotin in 1946, when the screwdriver separating two spheres of plutonium slipped and the material momentarily reached a critical state. He separated the pieces, but not before he was exposed to the radiation that killed him nine days later.[11]

There clearly were lethal limits of radiation that had to be avoided, though there was less certainty about what those limits were. The international and American committees initially adopted different tolerance figures and continued to make independent judgments in the years that followed. There was also debate over the notion of a threshold—a level below which radiation was less destructive. Was there such a level that simply had to be avoided? Or would cumulative exposure bring a result similar to a large harmful dose? The threshold approach had governed at first but was challenged in the 1940s. Several years after World War II, the American advisory committee, renamed the National Committee on Radiation Protection (NCRP), replaced the term "tolerance dose" with "maximum permissible dose" while reducing allowable limits for radiation workers; the international group, now called the International Commission on Radiation Protection (ICRP), took similar action. Agreement existed, according to Lauriston Tayler, still a member of the NCRP in the 1950s, that "any radiation exposure received by man must be accepted as harmful." But the concept of "maximum permissible dose" implied that social or political factors could be weighed against biological risks in making a final determination of what level was acceptable.[12]

Public concern about the effects of radiation mounted in the post–World War II years in response to the American nuclear-weapons testing program. The 1946 tests in the Pacific, two thousand miles southwest of Hawaii near the Bikini atoll, were the first to take place. American authorities, eager to experiment further with the bomb, sought a thinly populated, sheltered area where target vessels could be anchored and then destroyed. The Marshall Islands in Micronesia became the site for Operation Crossroads. After removing 167 Bikini inhabitants, the Navy brought in its own materials: 42,000 men, 242 ships, 156 planes, 25,000 devices to record radiation, and 5,400 experimental goats, pigs, and rats. Some less sophisticated observers feared that earthquakes or tidal waves might result, or that a blast might blow a hole in the bottom of the sea and let the water disappear. Others predicted that the island might be "atomized" and could then be renamed "Nothing Atoll." Those fears were unfounded as the explosions themselves proved anticlimactic. While the tests reassured military officials that the bomb could become an important part of the American arsenal, they dispelled initial anxieties about massive atomic destruction. The first test, its bomb carrying a picture of sex symbol Rita Hayworth, left the public most concerned about the fate of the experimental animals. California's San Fernando Valley Goat Association planned a service to remember the goats killed at Bikini, and *Life* magazine singled out surviving Pig 311, "the little animal that defied the big explosion," for special attention.[13]

The second test produced more ominous results. Not only was the blast—this time underwater—more impressive, but it created intense and unanticipated levels of radioactivity as well. Particles of fallout—residual radioactive droplets of water or dust—contaminated everything they touched. David Bradley, a physician and member of the radiological monitoring team, observed that reef fish, dried and placed on photographic plates, left radio-autographs or pictures of their internal architecture due to the radioactive substances they had absorbed. Ships entering the lagoon after the test were similarly tainted. In his book *No Place to Hide*, a journal chronicling his experiences in the Pacific, Bradley warned of "the invisible poison of radioactivity" that lingered after any blast. He argued persuasively that "there is no real defense against atomic weapons" and that "the devastating influence of the Bomb and its unborn relatives may affect the land and its wealth—and therefore its people—for centuries through the persistence of radioactivity." Condensed by *The Reader's Digest* and offered by the Book-of-the-Month Club, the book appeared on the *New York Times* best-seller list for ten weeks and sold a quarter of a million copies by the end of 1949.[14]

The public became more intimately involved with the testing program when the Truman administration selected a domestic site in 1950 to complement the Pacific locale. After examining spots in Utah, Nevada, New Mexico, and North Carolina, the AEC settled upon the Las Vegas–Tonopah Bombing and Gunnery Range, a 1,350 square-mile tract of uninhabited, government-owned land, larger than the state of Rhode Island, in the Nevada desert. Covered with brown grasses and rimmed by snow capped purple mountains, the new test site was located sixty-five miles northwest of Las Vegas and its twenty-five thousand inhabitants. Between 1951 and 1963, the United States conducted approximately a hundred above-ground tests there, with weapons usually detonated from weather balloons or towers.[15]

Most Americans were initially enthusiastic about the tests. Recognizing that public support of the program was necessary to ensure continued congressional funding, the AEC courted the national press. The commentary that emerged played up the spectacular side of the tests and ignored potential dangers. On April 22, 1952, a live television audience watched a bomb dropped from an airplane explode in midair. The visual impact was impressive enough, but the press, happy to serve as mediator, dramatized the detonations still further for the American people. A *Business Week* reporter, in a story entitled "Atomic Blast at Yucca Flat," noted: "The outstanding impression from first-hand observation of the explosion of an atomic bomb is this: How little there is in it of the horror and shock its historic importance would lead you to expect, how much of the sheer breathtaking beauty and magnificence." The next year, Chet Huntley, one of hundreds of reporters present at an atomic explosion, called the blast "the most tremendous thing I have ever experienced." Newspapers published schedules for atmospheric tests and directed interested parties to the best viewing sites. In Nevada, *Newsweek* declared, "those A-bombs are Las Vegas's alarm clocks." Utah residents often rose early to watch the atomic flashes and years later clearly recalled "the beautiful mushroom as it'd come up, [and] change colors." Californians could sometimes see the light. Millions of the nation's citizens witnessed the explosions directly, and many more followed them on television and in the press.[16]

Military personnel were among those most closely involved in the tests. Soon after the domestic testing program began, the Army insisted that soldiers be included, both to train them to deal comfortably with nuclear weapons and to monitor their responses. "Indoctrination in essential physical protective measures under simulated combat conditions, and observation of the psychological effects of an atomic explosion are reasons for this desired participation," the Pentagon's Military Liaison Committee told the AEC. To that end, the Army established Camp Desert Rock just outside the Yucca Flat site, to house military personnel from all branches. Over the next decade, approximately a quarter of a million servicemen participated in the tests. Positioned seven miles from the point of detonation initially, the men were later stationed much closer, eventually barely a mile away. Officers reassured them not to worry. "This is the greatest show on earth," Captain Harold Kinne told his forces in 1953 as they waited in trenches two miles from ground zero. "Relax and enjoy it." After the blast, the men were usually expected to charge closer, rifles and bayonets raised. Protective clothing was rare; often they were told simply to shower and change. Radiation limits were set, then raised, and seldom monitored closely.[17]

Despite repeated assurances from the AEC that there was nothing to fear, accidents and untoward incidents occurred, and the attendant publicity eroded public confidence. In 1955, the Defense Department acknowledged that four soldiers had suffered eye injuries from the brilliant flash of atomic blasts several years before. More troublesome was the fallout from the nuclear explosions. In March 1953, Dr. Lyle Borst, chairman of the Physics Department at the University of Utah and formerly a director of the AEC's Brookhaven National Laboratory, voiced his concern about the radioactive residue from ongoing atomic tests. He refused to let his children play outside and made them wash more frequently whenever a test occurred. "I would no more let my children be exposed to small amounts of radiation unnecessarily," he said, "than I would let them take small doses of arsenic." A month later, the Simon test in the Upshot-Knothole series left high levels of radioactivity in an area downwind from the site. Government officials checked vehicles and had to decontaminate a number of them.[18]

Worse still was the Harry shot that came to be known as "Dirty Harry." Exploded in May 1953, it blanketed residents of St. George, Utah, a small town east of the test site, with as much radiation as nuclear workers were allowed in a year. Local radio stations, following instructions from the AEC, warned people to stay indoors and to keep their windows closed for several hours. Several people in a town further east became ill, and forty-two hundred sheep grazing north of the test site died of mysterious causes. When ranchers claimed that fallout was responsible for the animals' deaths, the AEC launched an investigation. Examining bones, thyroid glands, and tissues, the agency concluded that radiation was not at fault. In 1956, the government successfully fought and won a suit filed by the ranchers on the grounds that the sheep owners had not provided scientific testimony to support their claim. Twenty-five years later, a federal judge ruled that the government had deliberately suppressed critical results and misrepresented the facts, and he called for a new trial.[19]

Countless Americans followed these episodes in stories in newspapers and the nation's most popular magazines. Even when the AEC and other agencies tried to put the best possible gloss on the situation, fears of possible consequences could not be contained. As criticism of domestic testing mounted, an accident in the Pacific dramatized continuing problems with fallout. On March 1, 1954, in the Bravo test, the United States exploded its first operational hydrogen bomb. Within hours of the blast, a snow-like ash fell on Rongelap—a small island a hundred miles east of Bikini—and its eighty-six residents. The islanders, who received twenty-five times the lifetime dose of radiation permitted thirty-five years later, soon became ill. Most had skin burns; many lost their hair. After measuring radiation levels, the United States evacuated the residents to Kwajalein atoll.[20]

Meanwhile, the crewmen of a Japanese fishing vessel, the *Fukuryu Maru* (*Lucky Dragon*), faced similar contamination. The tuna boat had been anchored ninety miles east of Bikini, well beyond the danger zone, when crew members saw a flash on the horizon and watched a red fireball rise into the sky. They felt, one of them recalled, "as if we were being chased by satans." They knew what had happened at Hiroshima and Nagasaki and understood they had experienced something similar, even if they were unaware that this blast was two hundred times more powerful. Yet it was not the explosion that harmed them; rather it was the fallout that winds carried toward the boat. Though the captain ordered the vessel to leave the area as quickly as possible, it was impossible to escape the falling radioactive ash. Before the end of the day, crewmen began to complain of radiation-sickness symptoms—headache, fatigue, nausea, diarrhea. Soon, they also suffered skin irritations and loss of hair. On their arrival in Japan two weeks later, they received medical treatment, but many remained ill. Seven months after the explosion, Aikichi Kuboyama, the ship's radio operator, died of jaundice, complicated by heart trouble and inflammation of the lungs. Whether or not hepatitis from a blood transfusion was the precipitate cause, his internal organs showed the pronounced effects of radiation that led to death. Worldwide attention focused on him as the first postwar victim of the nuclear age. Despite AEC chairman Lewis Strauss's contention that the *Lucky Dragon* was not a fishing boat at all but a "Red spy outfit" monitoring the tests, the American ambassador to Japan apologized to the Japanese, and the United States eventually provided two million dollars to compensate the Japanese fishing industry. But the damage was done. Fallout had become a serious international problem—in Ralph Lapp's words, "a peril to humanity."[21]

Some scientists in the 1950s were quick to highlight the increasing dangers of radiation. The American testing program, they suggested, harmed present and future generations and created an unacceptable human risk. In study after study, they charged that current policy could not be allowed to continue and provided a framework for other, less scientifically inclined critics to dramatize the dangers in more creative ways. Angered at the repeated contention of the AEC that there was nothing to fear, the scientists and their more creative supporters placed enough pressure on both the agency and the administration to effect first a temporary and then a more permanent change.

Many scientists, still reeling from the impact of the Oppenheimer hearing, welcomed the opportunity to speak out. They were often frustrated by what they considered the hostile attitude of the administration, dramatized by Secretary of Defense Charles E. Wilson's assertion that "basic research is when you don't know what you're doing." "American scientists remain, in 1956, a harassed profession, occupying a defensive position in the political arena," Eugene Rabinowitch, editor of the *Bulletin of the Atomic Scientists*, observed. "Their early hopes of playing an important role insuring world peace and prosperity are still in abeyance." The one scientist who was apparently heard was Edward Teller, and he remained out of favor with his colleagues after the Oppenheimer affair. The fallout issue gave others a chance to express their concerns once again.[22]

The debate over fallout, like the Oppenheimer case, polarized the scientific community. Sometimes scientists used different collections of data to support divergent views; on other occasions, they attached different conclusions to the same facts. The controversy stemmed from a still uncertain sense of the hazards of low-level radiation and then became intertwined with moral and political issues in assessing potential risk.[23]

The debate, carried out in public, began in early 1954, soon after the Bravo test. A. H. Sturtevant, a prominent geneticist at the California Institute of Technology, disagreed with Lewis Strauss's sanguine assessment of fallout damage and charged that radiation harmed both the exposed individuals and their descendants, since it affected somatic—or body—cells as well as germ—or reproductive—cells. There was, he said, "no possible escape from the conclusion that the bombs already exploded will ultimately produce numerous defective individuals," and he estimated that fallout was producing eighteen hundred mutations per year. The next year, Hermann Muller, a Nobel Prize–winning geneticist who had shown three decades earlier that radiation produced mutation, defended such quantitative estimates. A paper he submitted to a

U.N. conference on the peaceful uses of atomic energy was subsequently rejected at the instigation of the AEC, even though the agency had earlier accepted an abstract. Meanwhile, two other scientists, Ray Lanier and Theodore Puck of the University of Colorado, warned that radioactive fallout from a current domestic test series could threaten public health. They found themselves challenged by Colorado governor Edwin C. Johnson, a former senator and co-sponsor of the original bill to establish an atomic energy commission, who declared they "should be arrested."[24]

The scientific critique created pressure for an independent assessment of radiation hazards. In response, the National Academy of Sciences appointed several committees to investigate the problem, and their reports appeared in 1956. The Genetics Committee, which included Sturtevant and Muller, concluded that fallout had produced less radiation than various medical procedures but still counseled that exposure should be minimized since no amount of radiation was harmless. The Pathology Committee (which included several members with past or present ties to the AEC) proved more lenient as it defined an "unequivocally safe amount" of the radioactive isotope strontium-90 that humans could ingest. Ralph Lapp, increasingly concerned about fallout, argued that the Pathology Committee had disregarded the effects of an escalating arms race, and William Russell, principal geneticist at the Oak Ridge National Laboratory, charged that the committee's report contained major errors on issues settled long before.[25]

The following year, Linus Pauling, winner of a Nobel Prize in chemistry, estimated that ten thousand persons were either dead or dying of leukemia caused by nuclear tests. He also predicted that continued testing would lead to the birth of two hundred thousand physically or mentally defective children in each of the next twenty generations. One of his colleagues, biologist E. B. Lewis, demonstrated in an article in *Science* that a linear relationship existed between the incidence of leukemia and radiation dose. "It is apparent," the journal's editor commented, "that the atomic dice are loaded." A number of scientists testified in hearings on fallout before the Congress's Joint Committee on Atomic Energy in the spring and summer, and many more signed a petition, drafted by Pauling and others, calling for an end to nuclear tests.[26]

In the 1950s and early 1960s, popular culture came to reflect the scientific critique. Picking up on the fear of radiation that was becoming more widespread, commentators across the spectrum—authors, artists, filmmakers, and songwriters—let their imagination roam and produced a speculative but still devastating appraisal of radiation hazards and their consequences. Their efforts, like those of the scientists, forced the issue of fallout more squarely into the public arena and finally pressured policymakers into making a response.

A number of authors had fictionalized the ill effects of radiation even before the fallout issue surfaced. As early as 1947, in the short story "Tomorrow's Children," Poul Anderson and F. N. Waldrop told of a war that devastated the world and left a "damnable radiodust, blowing on the wind." Worst of all were the mutations—75 percent of all new births—that threatened to change the human race. That same year, in "The Figure," Edward Grendon described the consequences of the Alamogordo test on the insect world. Radiation enlarged the insects by about 40 percent and made them reproduce faster. Soon they surfaced by the millions. When scientists in the story produced a cube-shaped electrical field that operated in the future and showed the shape of things to come, they found a small statue astride a globe. The figure appeared intelligent and inspirational. "There is only one thing wrong," Grendon wrote in his final passage. "The figure is that of a beetle."[27]

Judith Merril explored similar themes with a more human focus the next year in her story "That Only a Mother," originally published, like "Tomorrow's Children" and "The Figure," in *Astounding Science Fiction*. She described Margaret, mother of a young baby, awaiting the return of her husband, Hank, who has never seen his child. Though Hank has worked around nuclear materials, Margaret tries desperately to persuade herself that he has not been affected. Even so, nightmares about genetic damage intrude: "*Stop it, Maggie stop it! The radiologist said Hank's job couldn't have exposed him. And the bombed area we drove past . . . No, no. Stop it, now!*" Margaret contents herself with taking care of her baby and writing glowing reports to Hank. When he finally arrives, he is stunned to see his daughter. Demanding an explanation, he realizes the horrible truth: "*She didn't know*. His hands, beyond control, ran up and down the soft-skinned baby body, the sinuous, limbless

body. *Oh God, dear God*—his head shook and his muscles con-
tracted in a bitter spasm of hysteria. His fingers tightened on his
child—*Oh God, she didn't know*."[28]

The film industry, likewise, had a field day with the subject
of radiation. Movies like *The Blob, The Attack of the Crab Monsters*,
and *Them!* all dealt with the effects of genetic damage caused by
radiation. *Them!*, which appeared in 1954, featured mutant ants
the size of buses crawling out of a New Mexico atomic test site.
The mutation, according to a scientist in the film, was "probably
caused by lingering radiation from the first atomic bomb." *Invasion
of the Body Snatchers* in 1956 showed pods from outer space landing
in southern California and taking over the minds of area residents.
At one point, the main character suggests that the relentless and
malignant disease could have come from atomic radiation. These
often financially successful films and others they inspired left viewers with the message that such creatures were the norm in the
atomic age.[29]

So, too, did the comic-book world confront the implications
of radioactivity. After World War II, Superman found himself
vulnerable to Kryptonite rays, as his authors now disclosed that
both he and Kryptonite came from a planet destroyed in an atomic
blast. In 1962, writer Stan Lee started a new comic-book story
featuring mild-mannered Bruce Banner, a scientist inadvertently
exposed to gamma rays in the test of a new bomb. Banner, "trembling on the brink of infinity," miraculously survives the blast but
is a changed man. Highly radioactive, he now has the ability to
change into "The Hulk." For Lee, mutation was not necessarily
harmful or evil. Bruce Banner was a mutant, to be sure, and a
potentially destructive mutant at that, yet he still deserved a measure of compassion. He was, Lee felt, a Frankenstein's monster
who "never wanted to hurt anyone" but "merely groped his tortuous
way through a second life trying to defend himself, trying to come
to terms with those who sought to destroy him." While he had
incredible power, it might be used for positive ends.[30]

In the same year he created "The Hulk," Lee introduced
another radioactive freak to comic-book readers. Peter Parker, "a
clean-cut, hard-working honor student" at Midtown High, is fascinated by the world of atomic science. In the course of a laboratory
experiment, a spider absorbs a large amount of radioactivity and

bites the closest living thing. Parker sees the burning, glowing creature on his hand and soon discovers that he now has incredible new powers. He has become "Spider-Man."[31]

Lee's cartoon characters were part of a world where frightening elements had uncertain effects. The characters may have been strange, but they were not evil; indeed, they often became forces for good. As a group, they reflected an attempt to come to terms with fears for the future, to show how radiation-induced changes might not be all bad. And yet, the notion of mutation still rankled. Despite efforts to show its possibilities, its potentially devastating effects could not be denied.[32]

The musical *L'il Abner* took a more frivolous view of the radioactive residue of atomic testing. Popular in the late 1950s, the comedy opens as Senator Jack S. Phogbound details the problems faced by the gambling industry in Las Vegas: "You see, in a desert nearby, the Government's a-shootin' off certain nuclear weapons of war. . . . And the atomic dust of them bombs is a'floatin' down an' a settlin' in the Las Vegas swimmin' pools, an' on them beautiful green felt crap-shootin' tables! . . . Ah tell you, the dice is a-turnin' black hidin' them spots so you can't hardly tell whether you got snake eyes or boxcars." To rectify this untoward situation, the government decides to find a truly worthless place, "the most unnecessary place in the whole U.S.A." to conduct its tests, and that place, of course, is L'il Abner's hometown of Dogpatch. The rest of the musical revolves around the question of whether Dogpatch will escape the ravages of the atomic tests.[33]

Songwriter Tom Lehrer was similarly satirical. The sometime mathematician, whose real genius lay in the songs he wrote and performed, included the consequences of fallout and atomic testing among his many targets. While still a graduate student at Harvard, he entertained friends with his raunchy songs and in 1953 released his first record. In "The Wild West Is Where I Want to Be," he caricatured the domestic testing program better than anyone else:

> Along the trail you'll find me lopin'
> Where the spaces are wide open,
> In the land of the old A.E.C.
> Where the scenery's attractive,

> And the air is radioactive,
> Oh, the wild west is where I want to be.

To deal with the obvious problems of fallout and radiation, the second verse suggested a solution:

> 'Mid the sagebrush and the cactus
> I'll watch the fellers practice
> Droppin' bombs through the clean desert breeze.
> I'll have on my sombrero,
> And of course I'll wear a pair o'
> Levis over my lead B.V.D.'s.

Six years later, on his second album, he took aim again at the atomic threat. Most concerned with the dramatic possibilities of atomic holocaust in "We Will All Go Together When We Go," he still managed to touch on the radioactive side effects of an extravagant blast:

> And we will all go together when we go,
> Ev'ry Hottentot and ev'ry Eskimo.
> When the air becomes uranious,
> We will all go simultaneous,
> Yes, we all will go together
> When we all go together,
> Yes, we all will go together when we go.

His black humor, here focused on a growing concern, first spread through a network of activists and social critics and finally reached audiences worldwide.[34]

In a more serious vein, folksinger Malvina Reynolds described the dangers of fallout in her haunting song "What Have They Done to the Rain?" In lines popularized by Joan Baez and others, she sadly observed the sickness and death riding on radioactive winds:

> Just a little boy standing in the rain,
> The gentle rain that falls for years,
> And the grass is gone, the boy
>   disappears,

And rain keeps falling like helpless tears,
And what have they done to the rain?[35]

Australian author Nevil Shute was equally pessimistic as he predicted the destruction of all human life from fallout in *On the Beach*. Published in 1957 and made into a popular American film in 1959, the novel told the story of a war that released so much radioactive waste that all life in the Northern Hemisphere was destroyed, while the Southern Hemisphere was reduced to waiting for the residue to come and bring the same deadly end. The conflict itself was a "short, bewildering war," a "war of which no history had been written or ever would be written now." But the real focus was on the aftermath of the struggle. Shute explored the creeping approach of the radioactive cloud by examining its imminent impact on an Australian couple with a baby and an American naval captain whose wife and family had already perished back home. There was no real hope in book or film, and both ended on a bleak note with nothing left. The final scene in the movie showed an empty town square. The message was clear: there would be no winners in nuclear war.[36]

The scientific critique, and the cultural criticism that stemmed from it, propelled the fallout issue into public view. "Talk and worry over the H-bomb's radioactive 'fall-out' is spreading," *Time* magazine reported in late 1954. The next year, *Science Newsletter* and the *Bulletin of the Atomic Scientists* explored the issue of whether tests should be suspended until the effects of radiation were better understood. *The New Republic, The Nation, The Reporter*, and *The Saturday Review of Literature* were among the publications calling for more information. "Are Atomic Tests Dangerous?" newsman Eric Sevareid asked. Those queries quickly made the public more cognizant of the problem. In the spring of 1955, a Gallup poll reported that only 17 percent of a national sample knew what fallout was; by the spring of 1957, 52 percent considered it a "real danger."[37]

As awareness increased, so did concern. Several mass-communications scholars noted in *Public Opinion Quarterly* a few years later that more information on the issue hardly proved reassuring. "Radioactive fallout itself," they reported, "is perceived as so devastating that when in addition there is a basic conflict among the scientists to whom one looks for authoritative clarification, it is small wonder that no reduction of anxiety was found despite knowledge, [or] media exposure." Near the end of the 1950s, the editors of *Playboy* voiced serious concern about strontium-90, which threatened their version of the good life. They observed, sensibly enough, that "these good things, this joy and fun, will cease to exist if life itself ceases to exist." At about the same time, journalist Steven Spencer described in *The Saturday Evening Post* the daily process of testing fallout levels. "No matter how you read it, the report is not good," he wrote. "Some say no weather report since the one given to Noah has carried such foreboding for the human race."[38]

Two studies contributed to the mounting public outcry. The first, conducted by the Committee for Nuclear Information at Washington University in St. Louis, examined how strontium-90 traveled up the food chain and ended up in baby teeth. The 1958 survey, which collected tens of thousands of tiny teeth, found alarming—and rising—concentrations of the radioactive isotope. A 1963 advertisement in the *New York Times*, showing three toothy grins over the caption "*Your* children's teeth contain Strontium-90," noted a sixteen-fold increase over five years.[39]

More troubling was the Consumers Union study of strontium-90 levels in milk. Applying its product-testing procedure to fallout, the organization surveyed milk samples in fifty different areas over a one-month period in 1958 and published the results in the March 1959 issue of *Consumer Reports*. "The Milk We Drink" was restrained but still frightening. Using graphs and charts, the organization noted that all readings fell below the National Committee on Radiation Protection limit yet revealed that the amount of strontium-90 in milk was rising. Its conclusion was far more pessimistic than normal in its product evaluations. "This report cannot be ended with a clear recommendation," it said. "None exists. No doubt the Best Buy is milk without strontium-90, air without fallout, and adequate medical care without diagnostic X rays. But none of these solutions are to be had, and it would be as foolish to stop drinking milk as it would be to refuse an X-ray examination for a broken limb." Despite that qualification, the article concluded that "there *is* a potential hazard" and called for more intensive government investigation of radiation problems and an end to nuclear testing.[40]

The public response was overwhelming. Newspapers around

the country printed banner headlines and feature stories about the milk tests. Medical journals and other specialized magazines noted the study. *Dog World* warned of the possible contamination of "the favorite canine food." Milk sales dropped.[41]

Faced with a growing public outcry about fallout, the AEC—and the administration—fought back. From the time the issue surfaced, the agency argued (without apparent contradiction) first that fallout was harmless and then that any possible danger was offset by knowledge gained from the tests. It took on the promotional task of reassuring Americans that past and present detonations were safe while at the same time persuading them that there was nothing to fear from taking reasonable risks. In 1953, in a semiannual report, it asserted that "the radioactivity released by fall-out has proved not to be hazardous." When Lyle Borst of the University of Utah complained that same year that it was neither "inconsequential nor trivial" to find as much contamination on his children as he had received in eighteen years of nuclear work, John Bugher, director of the AEC's Division of Biology and Medicine, responded for the agency. Sympathizing with Borst's concerns, he declared, nonetheless, that the AEC did not share the same philosophy about the danger of limited exposure to radioactive debris. Such exposure "should be kept as low as practical," he acknowledged, but even that restriction might be breached: "Atomic tests are conducted in the interests of national welfare, a circumstance which certainly warrants deviation from normal laboratory practices provided levels of radiation are kept well below levels of danger to the health of our personnel and the general population."[42]

The AEC clung to that position for the next several years. In 1955, it issued a small pamphlet on *Atomic Test Effects in the Nevada Test Site Region* that underscored the value of the Nevada blasts. "Because of them," it said, "we now have big bombs, and smaller ones too; in fact, a whole family of weapons," and it asserted that "an unusual safety record has been set," with no injuries and minimal property damage. It admitted that higher levels of fallout occurred when the fireball touched the ground but declared categorically that "no person in the nearby region has ever been exposed to hazardous amounts of radiation, even from this heavier fall-out, and no crops or water supplies have been made hazardous to health." Joining the public discussion about genetic damage, the pamphlet claimed that "radiation from fall-out from Nevada tests would have no greater effect on the human heredity process in the United States than would natural radiation in those parts of the Nation where normal levels are high."[43]

In agency discussions, the commissioners searched for ways to counter mounting criticism. When faced with complaints in 1955 from apple growers in Oregon and Washington that there were radioactive products in apples, chairman Strauss proposed making "a little catalog of these things to the extent that they turn out to be foolish, and eventually, perhaps not wait too long, put out a statement rebutting them in toto." Willard Libby, who shared Strauss's sense of the minimal dangers of fallout, took the lead in presenting the AEC's position. The only professional scientist then serving on the commission, Libby had worked on the Manhattan Project and had subsequently won a Nobel Prize in chemistry for a method of using radioactive carbon 14 to date fossils and other artifacts. Known as "Wild Bill," he was an apostle, according to *Time* magazine, of "bigger bombs and more bombs." Like Strauss, he was determined to do whatever was necessary to maintain the American testing program. In response to the concerns of the apple growers, he spelled out the agency's larger task: "Our job here is one of education. We must keep hammering at getting the facts to the people. If we relax on this educational task, we will pay by being bedeviled by such statements." Libby also noted that "one of our hazards in this is appearing to be stuffy and complacent and niggardly about information." In this particular instance, he counseled "infinite patience and good will" to make it clear that the matter was being carefully monitored. Over the next several years, Libby made speech after speech, highly regarded for their technical background even among those who disagreed with his basic premise that "people have got to learn to live with the facts of life, and part of the facts of life are fallout."[44]

The overall position of the Eisenhower administration was much the same. Civil defense authorities issued several million copies of pamphlets with titles like *Facts about Fallout* urging Americans not to panic when faced with residual radiation. When concern about fallout first surfaced, Ike himself relied on his astute sense of public relations to keep control of the situation. His initial approach was to encourage the AEC to exploit the public's lack of awareness. In mid-1953, then chairman Gordon Dean noted in his diary that Eisenhower had said, "Keep them confused as to 'fission' and 'fusion.' " In 1959, with the public more perceptive, he issued an executive order creating a Federal Radiation Council to placate critics by coordinating, in a non-binding way, the setting of radiation standards. The appearance of the film *On the Beach* that same year concerned top officials because of the impression of hopelessness the film conveyed. The Cabinet devoted considerable discussion to what it construed as " 'Ban the bomb' propaganda" and drew on AEC expertise to counter the movie's message at home and abroad. The theme, according to one background document, "is one of science fiction and not of scientific fact." The assumption that a war would wipe out all life was untrue, as was the image of people simply awaiting death passively. A message for U.S. Information Agency missions advised that "our attitude should be one of matter-of-fact interest, showing no special concern," but suggested that basic flaws perceived by the administration be shared with foreign officials and opinion leaders.[45]

Despite such efforts, perhaps because of them, protests against fallout continued, as did challenges to the AEC. Some years before, in dealing with the question of atomic radiation, Robert Sherwood, a Pulitzer Prize–winning playwright and former speech writer for FDR, had challenged "the theory that the way to keep the people fearless is to keep them ignorant," yet the perception persisted that the AEC was more concerned with maintaining secrecy and thereby kept pertinent facts about fallout from the public. In 1957, Michael Amrine, former managing editor of the *Bulletin of the Atomic Scientists*, charged the government with negligence: "Seldom has an American government withheld such vital information from its people, as in the case of the fall-out issue." Similar criticisms surfaced in other journals and hearings at home and abroad.[46]

Those criticisms were focused by the most visible protest group—the National Committee for a Sane Nuclear Policy, known simply as SANE. The organization had its origins in a meeting in mid-1957 called by Norman Cousins (*Saturday Review* editor) and Clarence Pickett (secretary emeritus of the American Friends Service Committee). Pickett set the agenda for the group of pacifists and non-pacifists who had responded to the invitation. "Something should be done," he said, "to bring out the latent sensitivity of the American people to the poisoning effect of nuclear bombs on international relations and on humanity." Psychologist Eric Fromm gave the group the name it eventually adopted when he declared, "We must . . . try to bring the voice of sanity to the people." SANE began its public activities with an advertisement placed in newspapers around the country that read: "We Are Facing A Danger Unlike Any Danger That Has Ever Existed . . . " A second ad proclaimed, "No Contamination without Representation." Within weeks, there was a groundswell of popular support, and by the end of the first year, SANE had about a hundred thirty chapters and twenty-five thousand members.[47]

SANE sought an immediate end to nuclear testing. It publicized the dangers from both blast and fallout in pamphlets with titles like *The Effects of Nuclear War*. More dramatically, it ran additional newspaper advertisements and created a number of television spots. A forty-second TV cartoon in 1958 told about *"The Bomb that Jack Built."* With the simplest art work, it pictured the bomb going "KABOOM," then showed the fallout from the bomb, the grass it contaminated, the cow that ate the grass, and the cow's milk that children drank. Equally effective a few years later was a full-page ad featuring internationally known pediatrician Benjamin Spock. Dressed in a suit with vest, he looked down at a little girl with a frown on his face. "Dr. Spock is worried," the caption read, with the text spelling out his concern. "I *am* worried," he said. "Not so much about the effect of past tests but at the prospect of endless future ones. As the tests multiply, so will the damage to children—here and around the world."[48]

A related group carried SANE's message still further. In 1961, five women who had been active in SANE grew restless at the male-led organization's emphasis on political lobbying rather than direct action. Meeting in the Washington, D.C., home of Dagmar Wilson—housewife, mother, and illustrator of children's books— they wanted to stress "mother's issues," like the radioactive contamination of milk. As one activist later noted, "This movement was inspired and motivated by mothers' love for children. . . . When they were putting their breakfasts on the table, they saw not only the Wheaties and milk, but they also saw Strontium 90 and Iodine 131." To protest continued testing, they called for women all over the country to suspend normal activities for a day and strike for peace. Word spread quickly around playgrounds, at PTA meetings, and through similar networks. On November 1, as a radioactive cloud from a recent Russian test drifted across the United States, an estimated fifty thousand women marched and mobilized in sixty communities around the country. They ran newspaper ads, offered radio and television interviews, and sought out local and national officials. Their slogans included "Let the Children Grow" and "End the Arms Race—Not the Human Race."[49]

Pressure from all groups finally brought a political breakthrough. In 1956, Democratic presidential candidate Adlai Stevenson, citing "the danger of poisoning the atmosphere," called for a halt to nuclear tests. Though Eisenhower was reluctant to discuss the issue in a political campaign, he let others counter Stevenson's charges. Vice President Richard Nixon declared that Stevenson was advocating "catastrophic nonsense," while Secretary of State John Foster Dulles dismissed the radiation hazard by saying, "From a health standpoint, there is greater danger from wearing a wrist watch with a luminous dial." Meanwhile, the President asked the National Security Council to consider once more the question of a test ban. When negotiations with the Russians broke down, the issue died, only to resurface after the election as the public became increasingly aware of the perils of radiation. Ike finally began to understand the need to address a worldwide outcry, particularly as the United Nations began to consider the question of fallout, rather than simply to respond to the military's call for continued tests. Thrown on the defensive when the Soviet Union announced a unilateral suspension of nuclear tests on March 31, 1958, the debate in the United States continued through another round of tests, with some scientists, Edward Teller among them, arguing passionately that any halt would undermine national security. They sought a "clean" bomb, with less fissionable fallout, as the answer to mounting public concern. Each time a breakthrough in negotiations appeared imminent, opponents came up with a counterargument. "These men who don't want a test moratorium are like a kid you are trying to put to bed," observed a member of the President's Science Advisory Committee. "First he wants a drink of water and then he wants to go to the bathroom, but what he really wants is not to go to bed." Finally, in the fall, the superpowers embarked on a voluntary moratorium that lasted until the Soviets resumed testing in September 1961; the Americans, the following March. That resumption, after John Kennedy had become President, provoked another wave of protest that paved the way for a more comprehensive agreement a few years later.[50]

Fallout focused atomic fears more than any other issue in the early postwar period. A silent and insidious killer that affected all people, and unborn children too, it dramatized the consequences of an arms race out of control. Americans and citizens of other nuclear powers around the world had to question whether the security of bigger and better bombs was worth the possibly corrosive trade-off. Still worried about a calamitous nuclear holocaust, which grew more likely as weapons improved, they now had to worry about a less violent form of extinction as well. Largely insulated from the earlier debates about atomic development and atomic control, the American public was very much involved in the debate over testing as the controversy over fallout became part of popular culture. Once aware of what might lie in store if atmospheric tests continued, the nation's—and the world's—inhabitants found it as difficult to dodge the issue as to avoid the radioactive residue itself. As scientists fought their battles in clear view, the dialogue shifted course. Scientific controversy sparked public concern and created pressure that government officials, whatever their efforts, could not ignore. Even attempts to orchestrate opinion fell short and failed to sidetrack the first agreements in the ongoing drive toward international control. The problem of fallout was not simply an extraneous element in the nuclear equation; it was an integral part of the extended debate.

5

Civil Defense

Was any protection possible from the ravages of nuclear war? That question preoccupied millions of people in the postwar era as civil defense became entangled with the larger issues of the atomic age. Americans who dreamed of a brave new atomic future worried at the same time about the terrifying consequences of nuclear war and pondered what kind of program might help if enemy bombs got through. Although concern about civil defense predated anxiety about fallout, the two issues became inextricably intertwined as the nation became better informed about the residual effects of a nuclear attack. Throughout the postwar period, particularly in the 1950s and 1960s, scientists and top officials debated whether any protective effort was warranted and, if so, what cost might be appropriate. Split among themselves, they succeeded in agitating the public and extending the debate but not in creating a supportive constituency, for critics, more often than not, had the upper hand in discussions about civil defense. Some of these critics acknowledged the need to do something but were unwilling to allocate the necessary funds. Others argued that the very fact of providing home-front protection might encourage a more aggressive military stance and thereby destabilize the already tenuous international balance. The most compelling opponents of civil defense claimed that the notion of protection from a nuclear bomb was one of the great myths of the atomic age and served only to distract Americans from more appropriate means of cooling tensions and minimizing the chance of attack. At a number of points during the Eisenhower and Kennedy administrations, there was intense interest in civil defense, and for a while shelter-building boomed. In the Reagan years, interest revived and civil defense experienced a resurgence of support. At no time, however, did a coherent national policy fully take hold and become a permanent part of American life. The cataclysmic fears fostered by proponents of protection simply helped strategists secure larger weapons budgets—with the argument that this was the only way to be entirely safe—while the civil-defense program remained a stepchild in the overall defense campaign.[1]

Scientists were among the first to sense the need for some sort of protection. Aware of the increasing force of the weapons they were helping to create in the postwar years, they shared their fears with those in policymaking positions, then watched as other Americans, in both public and private spheres, appropriated the issue of protection. While scientists continued to offer their advice about civil defense, the public proved to be an even greater source of pressure on the government. Fearful of fallout and worried about possible death, novelists and essayists, commentators and clerics argued vocally that something needed to be done. Officials in successive administrations then had to respond to that pressure or, as was more often the case, orchestrate it to their own ends.

Early demands to do something to safeguard the public drew on the civil-defense experience of World War II. The British had expended considerable effort in protecting themselves from German bombs, and their successful air-raid precautions had ensured their survival. Similar steps might well save large numbers of people in the nuclear age. In the United States, the Office of Civilian Defense (OCD) had similarly helped prepare for a wartime attack that never came, but its plans provided still another model for the postwar years.

The OCD had a checkered history. Created by executive order six months before the attack on Pearl Harbor, the agency had trouble establishing its own priorities. Fiorello La Guardia, the flamboyant mayor of New York City who served as first director, rejected shelters in favor of air-raid warning systems and warden schemes and similarly discouraged non-protective activities he regarded as "sissy stuff." When pressed to consider morale questions further, La Guardia finally appointed First Lady Eleanor Roosevelt assistant director, and she set up a series of national programs to encourage public participation in the defense effort. Her choice of actor Melvyn Douglas to lead a volunteer talent branch and of dancer Mayris Chaney to head a physical fitness section attracted the ridicule of political opponents who charged that the OCD was but another New Deal agency masquerading behind the needs of war. Resignations followed, and La Guardia was replaced by James M. Landis, dean of the Harvard Law School, who departed a year and a half later when it became clear that there was little probability of an attack at home. The agency, having done little more than cajole the public, was abolished in June 1945.[2]

Despite that demise, many Americans remained interested in civil defense as they recognized the awesome possibilities of the new atomic bomb. Worried at first about the implications of their own weapons, they became even more fearful of the bombs of the Soviet Union as a nuclear arms race began in 1949. Many agreed with English author C. P. Snow, who claimed a decade later that it was virtually certain that some bombs somewhere would go off, whether by accident or design.[3]

Although the OCD was dismantled in 1945, military planners continued to explore the possibilities of civil defense. United States Strategic Bombing Survey reports led the War Department to consider the issue more seriously, and at the end of 1946, Secretary of War Robert P. Patterson appointed a board to determine what defensive preparations might be made. Headed by Major General Harold R. Bull, the board concluded that "the fundamental principle of civil defense is self help." While it argued that civil defense was not a military responsibility, it acknowledged a military role even as it recommended that civilian agencies provide the necessary training of small groups in appropriate rescue skills.[4]

The next steps came a few years later. Early in 1948, Secretary of Defense James Forrestal established an Office of Civil Defense Planning and named Russell J. Hopley, a Nebraska telephone executive, as director in charge of formulating a national plan. The Hopley report suggested a small central office, with operational responsibility located in communities and states. At first reluctant to establish a permanent organization, President Truman gave civil-defense authority to the National Security Resources Board in 1949. The following spring, some months after the first Soviet atomic test, Congress's Joint Committee on Atomic Energy highlighted civil defense by holding public hearings on the question. As development of hydrogen weapons got under way, the need for protection seemed even more pressing, and on January 12, 1951, Truman established a new Federal Civil Defense Administration (FCDA).[5]

Despite creation of a bureaucratic structure, civil-defense efforts remained low key during the Truman years. Officials charged with responding to public pressure might agree that precautions were necessary, but the overall program continued to reflect its modest start, and the initiatives that were undertaken received neither the funding nor the support that proponents sought.

One such effort involved the dispersion of the urban and industrial capacity of the United States. As early as 1946, Edward Teller and two collabotators argued in the *Bulletin of the Atomic Scientists* that "a country like the United States with a large part of its population concentrated in big cities along the eastern seaboard is particularly vulnerable to the devastating impact of atomic bombs." Science-fiction author Robert Heinlein made a similar point in the early postwar period when he shifted briefly from writing imaginative stories and sought to alert his readers to "the *meaning* of atomic weapons." Decentralization was essential, he wrote in "The Last Days of the United States," whatever the cost: "The cities must go. Only villages must remain. If we are to rely on dispersion as a defense in the Atomic Age, then we must spread ourselves out so thin that the enemy cannot possibly destroy us with one bingo barrage, so thin that we will be too expensive and too difficult to destroy." Without being quite so cataclysmic, the United States Strategic Bombing Survey underscored the value of decentralization as it pointed out that factories and other installations on the periphery of both Hiroshima and Nagasaki escaped major damage.[6]

Proponents picked up on the theme in the next several years. In 1948, Tracy B. Augur, a prominent city planner, declared that the cost and inconvenience of dispersion could be minimized by integrating it with current building plans. He proposed rechanneling urban growth into a network of smaller cities, with small "clusters" of buildings separated by bands of open space. Two years later, physicist Ralph Lapp agreed that "dispersion is the only really effective answer to the atomic bomb." In September 1951, staking out its own support in an editorial entitled "The Only Real Defense," the *Bulletin of the Atomic Scientists* devoted an entire issue to the question.[7]

As pressure mounted, Truman announced a National Industrial Dispersion Policy. A 1951 executive order required that federal aid for new industrial plants be conditioned on their placement in invulnerable sites. By the time the Truman administration was ready to leave office, a report indicated that the program, largely voluntary, was successful in encouraging new defense plants to locate in less congested areas. But an effort to disperse federal government buildings had failed.[8]

A shelter program initiated during Truman's incumbency proved equally limited. Some of those pushing for civil defense favored evacuation in the event of attack, while others promoted shelters instead. Although the shelter advocates won out, they were soon impaled on the problem of cost. Before fallout became a concern, planners were preoccupied with blast-proof structures, which were prohibitively expensive. The first estimates for a nationwide program ran to thirty-two billion dollars over a five-year period. Even a scaled down figure of sixteen billion dollars was far too high for officials anxious to keep the budget under control.[9]

Despite predicted costs, interest in shelters remained strong. In California, Governor Earl Warren urged the building of family shelters. The *Chicago Tribune* told readers that publisher Robert McCormick had already built one for himself. In Washington, D.C., a $938,000 bomb shelter was planned for the White House, and Missouri officials proposed using a planned underground garage in Kansas City, so that if an attack came, "we would all run over there from the little White House with a case of ginger ale and some biscuits and camp out until the smoke blew away." Even before the creation of the Federal Civil Defense Administration, other agencies began work on designs for windowless buildings and considered the pros and cons of basement and garage shelters. They also authorized studies to develop data on which a more coherent program could be based. Recognizing the impracticality of a large-scale effort, they were, nonetheless, unable to accept the notion of no shelters at all. The order creating the FCDA recommended a shelter program with a tab far lower than initially proposed, to be shared by the federal government, the localities, and the states.[10]

Authorization was one thing; implementation was another. At no time during the Truman years was the government willing to spend what advocates asked. In 1951, for example, Congress took a $535 million request for civil defense and cut it to $75 million, eliminating shelter funding entirely. Similar cuts in succeeding years perpetuated that pattern.[11]

With but limited funds, civil-defense activity necessarily involved a good deal of cajoling. Pamphlets, films, television shows, and other displays all aimed at showing the American public how to cope with an atomic attack. *Survival*, a seven-part television series shown on NBC in 1951, reached an estimated twelve million people. A film, *Survival Under Atomic Attack*, was made for civil-defense authorities and produced under government direction with private money. Narrated by news commentator Edward R. Murrow, it circulated throughout the country with conspicuous success. A widely circulated booklet with the same title told readers on the first page: "You can SURVIVE," and went on: "You can live through an atom bomb raid and you won't have to have a Geiger counter, protective clothing, or special training in order to do it. The secrets of survival are: KNOW THE BOMB'S TRUE DANGERS. KNOW THE STEPS YOU CAN TAKE TO ESCAPE THEM." The pamphlet counseled people to drop to the ground or floor and shield themselves as much as possible.[12]

The "Alert America" campaign that began at the end of 1951 conveyed a similar message. Three convoys of ten 32-foot trailers each carried portable exhibits to cities throughout the United States. They contained dioramas showing the possible impact of a bomb on a typical city and posters showing how civil defense could help. "TRAINED, ALERT CIVILIANS will cut our casualties in half," one declared. "IF WE ARE PREPARED," another said, "WE CAN COME BACK *fighting*." All materials attempted to arouse public involvement and concern.[13]

Other efforts were aimed at school children. Educational systems around the country initiated atomic air-raid drills to prepare students for a possible attack. In the year following August 1950, schools in New York City, Philadelphia, Chicago, Detroit, Milwaukee, Fort Worth, San Francisco, and Los Angeles all started such exercises. Los Angeles teachers developed surprise drills, in which they shouted "Drop!"; the children responded by kneeling down, hands clasped behind their necks. New York City public schools developed "sneak attack drills," even though those had not yet been mandated by the state Commission on Civil Defense. Meanwhile, educators began to consider the problem of identification. Modeling metal tags after military dog tags, New York City led the way and by April 1952 had issued two and a half million free tags to all public, private, and parochial school children. Other cities followed suit. San Francisco provided free tags; Philadelphia sold them in local stores; Seattle arranged for their distribution through the PTA. Though advocates understood that the tags would be most valuable in identifying victims, they also implied that in some mystical way the markers could help protect Americans in a nuclear attack.[14]

That same naive optimism infused the "Duck and Cover" campaign targeting school children. In three million comic books distributed nationwide, Bert the Turtle stressed the need to take cover from flying glass and other debris in case of a raid. One frame told young readers that in the face of danger "BERT DUCKS AND COVERS. HE'S SMART, BUT HE HAS HIS SHELTER ON HIS BACK. YOU MUST LEARN TO FIND SHELTER." A subsequent frame advised. "OUTDOORS, DUCK BEHIND WALLS AND TREES. EVEN IN A HOLLOW IN THE GROUND. IN A BUS OR AUTO, DUCK DOWN BEHIND OR UNDER THE SEATS." And a final one concluded, "DO IT INSTANTLY. . . . DON'T STAND AND LOOK. DUCK AND COVER!" Bert the Turtle also starred in an animated film that brought the same message to children who may have missed the comic. A lively jingle in the film helped make the point:

> There was a turtle by the name of Bert.
> And Bert the Turtle was very alert.
> When danger threatened him he never got hurt.
> He knew just what to do.
>
> He'd Duck and Cover. Duck and Cover.
> He did what we all must learn to do.

> You and you and you and you.
> Duck and Cover!

The cheery enthusiasm of the comic book and film made nuclear war seem like a minor inconvenience that could be handled playfully without undue disruption.[15]

Americans did become better informed about how they might deal with nuclear raids in the late 1940s and early 1950s. As they learned to live with the bomb, many of them began to accept Truman's assertion that "there is no complete protection against an atomic air attack, but there is a great deal that can be done to reduce the number of deaths and injuries that might result." Wanting to do something to avoid the consequences of a possible strike, they responded good-naturedly to the cajoling they received from local and national authorities. But even though they participated in the public campaigns the administration endorsed, they were still reluctant to push too hard for the kind of comprehensive program proponents demanded. As FCDA administrator Millard Caldwell observed, too many people simply assumed that the nation's armed forces could fend off any assault and were therefore less concerned about protection. "We have made progress," he concluded, "but far from enough."[16]

The stakes rose in the Eisenhower era. Nuclear tests, played out in public view, exacerbated fears and encouraged further consideration of civil defense. Demonstrations in 1953 vividly dramatized the power of an atomic blast. Both *Time* and *Newsweek* ran sequential pictures of a frame house in a simulated suburb disappearing into a pile of rubble. In the space of a just over two seconds, the structure was first lit, then engulfed in flames, and finally blown to bits. It looked, *Time* noted, "like a match box crumpled on a table." Mannequins in peaceful domestic scenes were splayed in contorted positions after the explosion. Their fate offered graphic evidence of the likely human consequences of an attack. Three quarters of the American people reported in a survey that they had heard about the burst or seen it on TV and were now even more aware of the bomb's destructive force. Two years later, administration officials sparked further fears as they commented on the Bravo test of a hydrogen bomb virtually ready for delivery. President Eisenhower acknowledged that even scientists were astonished at the weapon's power, and AEC chairman Lewis

Strauss left a press conference aghast with his admission that "an H-bomb can be made . . . large enough to take out a city," even New York. The discovery of the danger of fallout only sharpened public concern.[17]

The development of the hydrogen bomb further undermined the first postwar protective-policy efforts and led to demands for a different approach. "The present national dispersion policy is inadequate in view of existing thermonuclear weapons effects," the Atomic Energy Commission acknowledged in late 1954. Likewise, the shelter program was obsolete. "The advent of the H-bomb," an FCDA official wrote the next year, "has necessitated a modification in earlier civil defense planning. The only 100% defense against it is not to be there when it goes off." Val Peterson, agency head between 1953 and 1957, was even more pointed: "The alternatives are to dig, die, or get out; and certainly we don't want to die." Digging seemed far too costly in the face of ever more potent bombs, so the Eisenhower administration's approach became one of evacuation if attacked. The orientation changed, according to the *Bulletin of the Atomic Scientists*, "from 'Duck and Cover' to 'Run Like Hell.' "[18]

But evacuation required an adequate system of roads. The 1956 act creating an interstate highway system provided easier automobile access to the suburbs and, at the same time, an expeditious means of exit from the cities in case of nuclear war. Eisenhower was proud of the enormous project. In his memoirs, he waxed eloquent about the massive amounts of concrete poured that could have made "a parking lot big enough to hold two thirds of all the automobiles in the United States" or "six sidewalks to the moon." This project, more than any other since the war, "would change the face of America. . . . And motorists by the millions would read a primary purpose in the signs that would sprout up alongside the pavement: 'In the event of an enemy attack, this road will be closed.' "[19]

While planning the highway system, the administration began to check the readiness plans of various government agencies in case of attack. A three-day "Operation Alert" in June 1955 sought to determine if officials could evacuate Washington and perform regular and emergency functions at relocation sites. The exercise began with a hypothetical raid on Washington and fifty-four other cities that theoretically left about eight million people dead, seven million injured, and twenty-four million homeless. The President and a host of other officials left Washington for retreats outside the city, where they attempted to deal with the crisis. Despite an optimistic report of "an exceptionally fine job," the exercise revealed serious problems. Thousands of people regarded the drill as useless and simply ignored it. Even top officials occasionally followed their own bent. When the Secretary of Health and Education, Oveta Culp Hobby, showed up late at her site and was asked where she had been, she confessed that she had stopped for lunch along the way. In Washington, John Garrett Underhill, formerly a military intelligence officer and now a civil-defense worker, called the test "a fiasco" and charged that it was "so inadequate it couldn't cope with a brush fire threatening a doghouse in the backyard." Somewhat more charitably, *Newsweek* concluded that "Operation Alert had proved one thing: The nation still was not ready to cope with a nuclear attack." Unwilling to dispense with the relocation program, the government conducted similar tests in succeeding years.[20]

The whole notion of evacuation, however, suffered a fatal blow with the growing realization of the consequences of fallout. The creeping radioactive cloud that accompanied any nuclear blast minimized the value of running away, for the insidious dust remained deadly without some shield. Blast shelters remained useless to safeguard people from a bomb's impact, but less substantial and less costly fallout shelters might, nonetheless, save their occupants from the poisonous side effects of an attack. Hearings before the House Military Operations Subcommittee of the Committee on Government Operations that began in early 1956 asserted the need for a new national effort. Headed by Congressman Chet Holifield, the subcommittee took testimony from doctors, scientists, engineers, and public officials for the next year. Concluding that the current civil-defense program was not working at all, it proposed greater authority for administrators, development of a master plan, and a nationwide fallout-shelter program to be financed largely with federal funds.[21]

As the hearings unfolded, the administration had to respond to the agitation for a more effective form of civil defense. Eisenhower himself was increasingly worried about the prospects of surviving a nuclear war. In 1955, after a Pentagon briefing about the consequences of an attack on military targets alone, he noted that recovery "would literally be a business of digging ourselves out of the ashes, starting again." The following year he asked members of the National Security Council to consider what to do when "we reach a point where we will have passed the limits of what human beings can endure." Val Peterson's proposal for a far-reaching shelter program that could have cost between twenty billion and forty billion dollars was considerably more than the President had anticipated. Taken aback by the massive request from his normally conservative civil-defense director, Eisenhower appointed a committee of private citizens to study the problem. Chaired by H. Rowan Gaither, Jr., California attorney and board chairman of both the Ford Foundation and the RAND Corporation, the Security Resources Panel had a limited mandate at the start. Yet the question Ike asked when he assembled the group at the White House was more open-ended. "If you make the assumption that there is going to be a nuclear war," he wondered aloud, "what should I do?" Taking the President's question at face value, the panel expanded its inquiry on its own initiative to include a variety of defense issues.[22]

The Gaither committee "grew like a cancer," according to one participant, and eventually involved nearly a hundred experts. In the group were scientists from the Manhattan Project, like I. I. Rabi and Herbert York, and such strategists as Paul Nitze and Albert Wohlstetter. The committee's expanded mandate led it to consider protection of bombers and missiles as well as citizens and to contemplate the kinds of weapons the military should buy. In its report, titled "Deterrence and Survival in the Nuclear Age" and given to the President in late 1957, the panel pictured a nation in danger. "If we fail to act at once," it declared, "the risk, in our opinion, will be unacceptable." With inflammatory language, it focused on the threat posed by the surprisingly early Soviet development of a successful intercontinental ballistic missile (ICBM) and recommended protective steps to safeguard the nation's retaliatory forces and its people, including a substantial fallout-shelter program. The price tag on such steps might reach fifty billion dollars over the next few years. The report, issued soon after the Soviet launch of *Sputnik I* and subsequently leaked to the press, frightened those who faced difficult decisions. "It was like looking into the abyss and seeing Hell at the bottom," according to former Secretary of Defense Robert Lovett.[23]

Eisenhower was not persuaded by the report. "Hell, these people come in here and they tell me things I've known all along," he declared. "I'm not going to dance at the end of the string of . . . people who try to give me . . . scare stories." On the issue of civil defense that prompted the study, he shared the reservations of Secretary of State John Foster Dulles. At the meeting of the National Security Council during which the Gaither panel report was presented, Dulles immediately objected that the Soviet Union would misunderstand a large shelter program and observed that allies who could not afford similar protection would resent it. Later, when Leo A. Hoegh, formerly governor of Iowa and now civil-defense administrator, argued that a shelter program costing $22.5 billion could save fifty million American lives, Dulles became even more critical. "If a wave of a hand could create those shelters," he said, we'd of course be better off with them than without them. But it's hard to sustain simultaneously an offensive and defensive mood in a population. For our security, we have been relying above all on our capacity for retaliation. From this policy we should not deviate now. To do so would imply we are turning to a 'fortress America' concept." Though Ike commented to Dulles, "You *are* a militant Presbyterian, aren't you?" he agreed with the Secretary's basic position and chose not to embark on a massive shelter program.[24]

Instead, Eisenhower followed his penchant for cajoling people to take action on their own. In May 1958, he issued a National Shelter Policy, which operated on the assumption that every citizen was responsible for his or her own protection, and emphasized private shelter construction by homeowners. The federal government would provide leadership, direction, and advice, but little funding. In August, after a bureaucratic reorganization created a new Office of Civil and Defense Mobilization (OCDM), he signed another measure authorizing federal financial assistance for state projects, but the money allocated was minimal. Civil-defense requests for the last half of the decade averaged a mere $102.7 million annually; appropriations averaged an even more miniscule $60 million.[25]

Nonetheless, public attention remained riveted on shelters. Americans inclined to act debated the merits of different kinds of construction. Edward Teller, always eager to join in the latest controversy, continued to advocate a chain of deep underground shelters, each able to hold a thousand persons, as the best defense from the ravages of attack, but most other participants in the debate favored more modest structures that could protect smaller groups of people from fallout alone. The federal government encouraged interest by building prototype shelters around the country and issuing a number of pamphlets to encourage citizens to follow that lead. The OCDM distributed free copies of *Family Fallout Shelter,* a booklet that told readers how to protect themselves. Another government publication told *How It Was Done* in rural Gratiot County, Michigan. It featured Sheriff Bob Russell as he constructed a fallout–tornado shelter in his new house. Russell declared that "to build a new home in this day and age without including such an obvious necessity as a fallout shelter would be like leaving out the bathroom 20 years ago." The Boy Scouts of America delivered forty-two million copies of still another civil-defense publication—an emergency manual—to homes throughout the United States.[26]

Even more important in stimulating public support was the popular press. *Good Housekeeping* magazine in November 1958, for example, carried a full-page editorial that encouraged the building of family shelters. When it urged its readers to write the OCDM for free copies of the *Family Fallout Shelter* booklet, fifty thousand people responded. The *Shawnee News-Star* in Oklahoma noted that the need to build shelters was a sad commentary on the human condition but still concluded that preparation for any contingency was necessary: "We laud the thrifty squirrel that gathers food in the summer to carry it through the long winter. We scorn the silly butterfly which does nothing but enjoy the summer months without preparation for survival when winter comes." A cartoon in Pennsylvania's *Harrisburg News* showed Noah's ark with the caption, "They laughed at Noah!"[27]

Shelter-building came of age in the Eisenhower years. *Life* magazine reported in 1951 on backyard shelters in California that ran the gamut from an eight-dollar wood-covered dirt hole to a fifty-five-hundred-dollar complex with telephone, radio, water supply, and Geiger counters inside and outside. It also pictured a sixty-five-dollar septic tank that could be converted into a six-person shelter. While there were no takers for that makeshift structure, one contractor noted orders for fifty-four shelters costing between five hundred and eighteen hundred dollars apiece. "For those who can afford neuroses," another builder declared, "this is it." With the advent of new weapons, *Life* returned to the issue of protection by featuring an "H-Bomb Hideaway" for three thousand dollars. As fallout caused increasing concern, interest in shelters became even more widespread. In 1959, a Miami firm reported numerous inquiries about shelters that sold for between $1,795 and $3,895, depending on capacity, and planned nine hundred franchises. By the end of 1960, the OCDM estimated that there were a million family fallout shelters nationwide.[28]

Shelters large and small became part of the cultural landscape in the 1950s. Creative construction patterns were important, as the administration attempted to persuade women to participate in the civil-defense effort. At the end of the decade, the government asked the American Institute of Decorators to design a shelter that could look inviting enough for everyday living. Built properly, a shelter might double as a playroom or a family den. Jean Wood Fuller, an FCDA official who had served as a "female guinea pig" in a 1955 Nevada test, worked tirelessly to promote a program of "home protection and safety." With assistance from the National Grocer's Association and the American National Dietetic Association, she launched the "Grandma's Pantry" campaign to encourage home bomb shelters. "Grandma's pantry was always ready," a government brochure declared. "She was ready when the preacher came on Sunday or she was ready when the relatives arrived from Nebraska. Grandma's Pantry was ready—Is Your Pantry Ready in Event of Emergency?" A shelter, women were told, might also serve as more than a pantry. Author Betty Friedan referred caustically in *The Feminine Mystique* to an article submitted to a leading women's magazine entitled "How to Have a Baby in an Atom Bomb Shelter." Despite her outrage at the assumption that "women, in their mysterious femininity, might be interested in the concrete biological details of having a baby in a bomb shelter, but never in the abstract idea of the bomb's power to destroy the human race," the promotional campaigns aimed at women continued.[29]

In short, civil defense became something of a national preoccupation. An entire generation came of age taking shelters, and the drills that promised supplementary protection, for granted. Years later, Vietnam veteran Ron Kovic fondly recalled growing up on Long Island, where he and his friends "made contingency plans for the cold war and built fallout shelters out of milk cartons."

Advertising copywriter Leslie J. Miller, thinking back to the 1950s, observed that "I wasn't the only one who begged her parents for a shelter." She went on to describe the experience of another girl who bought candy for one of the "nerdiest girls in the class." Why? "Because *her* parents had a fallout shelter." School drills sat children down in hallways or under their desks, as the rooms and corridors became massive shelters. Author Annie Dillard, remembering her childhood in Pittsburgh, captured the light-hearted optimism that marked the exercises in her school: "We tucked against the walls and lockers: dozens of clean girls wearing green jumpers, green knee socks, and pink-soled white bucks. We folded our skinny arms over our heads, and raised to the enemy a clatter of gold scarab bracelets and gold bangle bracelets." If the bomb came, the older children were prepared to help the younger ones: "We would help them keep spirits up; we would sing 'Frere Jacques,' or play Buzz." Shelter *was* important, millions of Americans believed, whether found in a basement hideaway or a school hall. With proper preparation, they could survive an attack in modest comfort, without serious disruption to their lives.[30]

    Yet there was a lurking underside to that naive public mood. Shelters themselves might not withstand an atomic blast, and their vulnerability led to lingering qualms. In a 1957 Nevada test, an aluminum structure simply melted from the heat of the explosion and revealed that some alternative form of protection was necessary. But even shelters remaining intact might not ensure survival. In *No Place to Hide*, a thirty-minute documentary made in the early 1980s, filmmakers Tom Johnson and Lance Bird focused on the generation raised with the assumption that civil defense made sense and concluded that, despite official promises of safety, many people worried that there really was no escape from atomic attack. Novelist Tim O'Brien, part of that generation, worked through his own anxieties years later in fictional form and captured the haunting fears he could not avoid:

> When I was a kid, . . . I converted my Ping-Pong table into a fallout shelter. Funny? Poignant? A nifty comment on the modern age? Well, let me tell you something. The year was 1958 and I was scared. Who knows how it started? Maybe it was all that CONELRAD stuff on the radio, tests of the Emergency Broadcast System, pictures of H-bombs in *Life* magazine, strontium 90 in the

milk, the times in school when we'd crawl under our desks and cover our heads in practice for the real thing. Or maybe it was rooted deep inside me. In my own inherited fears, in the genes, in a coded conviction that the world wasn't safe for human life.[31]

Popular literature in the 1950s reflected the ambivalence of the period. Two novels at the end of the decade explored the likelihood of life after nuclear war and came to different conclusions. In 1959, after corresponding with FCDA authorites, Pat Frank assumed that survival might be possible after all. In *Alas, Babylon*, he described Fort Repose in central Florida after a nuclear attack. The town itself was not destroyed, yet "after The Day, the character of Fort Repose had changed. Every building still stood, no brick had been displaced, yet all was altered, especially the people." Frank's characters struggle successfully to re-create a functioning society. They deal with highwaymen, reestablish lines of authority, find new sources of food, and display incredible ingenuity in their struggle to survive. Life goes on, as they discover "that faith had not died under the bombs and missiles." Their job was not easy, Frank acknowledged, but they can make it in the end.[32]

At about the same time, Walter M. Miller, Jr., took a broader and ultimately more discouraging view. In *A Canticle for Leibowitz*, which appeared first as three stories in *Fantasy and Science Fiction* between 1955 and 1957 and then in book form in 1959, he told the tale of the rise and fall of civilization as humans failed to learn the lessons of their deeds. Miller's book, which became something of a classic, remained in print for years, and was occasionally read aloud in its entirety on the radio by anti-nuclear activists, described the world after "the Flame Deluge." In a monstrous nuclear cataclysm, "cities had become puddles of glass, surrounded by vast acreages of broken stone." The land remains "littered with bodies, both men and cattle, and all manner of beasts," though some survive. Working far more slowly than the survivors in *Alas, Babylon*, they begin to re-create some semblance of society, but it takes centuries to achieve even the smallest gains. Examining the interplay between science and religion, Miller painted an agonizing picture of the rebirth of civilization, only to recount the descent into oblivion once more. "Are we doomed to do it again and again and again?" the Abbot Dom Zerchi asks himself. "Have we no choice but to play the Phoenix in an unending sequence of rise and fall?" Though there were elements of hope in this complex novel, especially as a spaceship takes off for the stars at the end, Miller's message was far more ambiguous than Pat Frank's as he reflected on the possibilities of sustaining meaningful human life in the nuclear age.[33]

The hopes—and fears—about the prospects for survival continued into the early 1960s. A change in administration led to even greater concern with safeguarding the population of the United States. John F. Kennedy entered office in early 1961 intent on giving new direction to the nation and providing more fully for its defense. Civil defense was part of his plan.

Kennedy's own convictions guided his defense policy. The youngest man ever elected to the White House, he stressed the importance of strong leadership and believed that the President must be a catalyst who was "prepared to exercise the fullest powers of his office—all that are specified and some that are not." He was especially interested in providing that kind of direction in the field of foreign affairs. He shared the Cold War assumptions of his precedessors and was determined to stand firm in the face of any threats from abroad. During the 1960 campaign, at the Mormon Tabernacle in Salt Lake City, he had declared, "The enemy is the communist system itself—implacable, insatiable, unceasing in its drive for world domination." His rhetoric, as both campaigner and President, was soaring but shrill. The United States, he proclaimed in his inaugural address, would "pay any price, bear any burden, meet any hardship, support any friend, oppose any foe to assure the survival and the success of liberty."[34]

During its first months in power, the new administration devoted considerable attention to questions of defense. It completed plans to drive Fidel Castro from Cuba that culminated in the abortive invasion at the Bay of Pigs. Though its campaign charges of a missile gap—in which the Soviet Union had a clear advantage over the United States—proved unfounded, it chose, nonetheless, to build a larger missile force to guard against future Soviet expansion. Meanwhile, it examined past policies and proposals in the area of civil defense and determined to push even harder for a coherent program. Kennedy felt pressured by New York governor Nelson Rockefeller, a vigorous proponent of shelters who many advisers thought would be the Republican candidate for President in 1964. Despite staff dissent, Kennedy concluded he needed to do something. "One major element of the national security program which this nation has never squarely faced up to is civil defense," he declared in May of his first year. "This problem arises not from present trends but from national inaction in which most of us have participated." Fallout protection was essential, and shelters provided necessary insurance "which we could never forgive ourselves for foregoing in the event of catastrophe." Like Eisenhower, Kennedy encouraged individual activity, but he also indicated that increased federal funding would follow.[35]

Two months later, a foreign crisis highlighted the issue. Berlin had long been a major area of superpower confrontation in the Cold War. Germany and Berlin itself had been divided into zones at the end of World War II, and the split had eventually resulted in two nations—East and West Germany—and two city sectors—East and West Berlin. A Soviet blockade of roads and waterways leading to West Berlin had culminated in the Berlin airlift in 1948; tension had recurred in 1958, when the Russians challenged the postwar system but then backed off again. As Kennedy assumed office, Soviet leader Nikita Khrushchev demanded once more an end to occupation arrangements and insisted on a permanent settlement soon. Afraid that Khrushchev's demand reflected larger Soviet designs, Kennedy refused to give ground. Though he doubted the usefulness of summit conferences, he journeyed to Vienna in June to meet with Khrushchev and was bullied by the Russian leader. Khrushchev took the initiative, asserting his intention of supporting "wars of national liberation" and even threatening war over Berlin. "It will be a cold winter," Kennedy predicted as he left the meeting.[36]

Aware of the need to avoid the evils of "holocaust or humiliation," Kennedy still resolved to stand firm, even at the risk of sparking nuclear war. Impulsive, insecure about his paper-thin electoral margin, and captive of the rhetoric that promised a sense of purpose, he seized the initiative from his more cautious State Department. On July 25, he addressed the American public on television, in what international-affairs analyst Michael Mandelbaum later termed "one of the most alarming speeches by an American President in the whole, nerve-wracking course of the Cold War." Berlin, Kennedy said, was "the great testing place of Western courage and will." He asked Congress for an increased defense

appropriation of more than three billion dollars and called for
doubling draft calls, activating reserve units, and bolstering the
size of the armed forces by more than two hundred thousand men.
At the same time, he transferred civil-defense responsibilities to
the Pentagon where closer coordination with military planning was
possible.[37]

He also called for an expanded shelter program. "To recognize
the possibilities of nuclear war in the missile age, without our
citizens knowing what they should do and where they should go
if bombs begin to fall," he declared, "would be a failure of responsibility." He asked for—and received—$207.6 million for fall-
out shelters, with most of the money allocated for marking and
stocking existing community shelter spaces in subways, tunnels,
corridors, and basements that could temporarily accommodate fifty
million Americans. The President's grim assessment of the international situation was not lost on the American public, and his
plea for a comprehensive civil-defense program in a time of crisis
gave the issue even greater credibility than it had before.[38]

The administration undertook a multifaceted civil-defense effort. In September, *Life* magazine carried a message from the
President which told citizens that "there is much that you can do
to protect yourself—and in doing so strengthen your nation." The
article on fallout shelters that followed underscored his point: "You
could be among the 97% to survive if you follow advice on these
pages ... How to build shelters ... Where to hide in cities ...
What to do during an attack." Taking his earlier initiative as a
starting point, Kennedy proposed a five-year shelter program to
protect the entire population. Such an approach, he told Richard
Russell, chairman of the Senate Armed Services Committee, would
"greatly increase the capacity of this country to survive and recover
after a nuclear blow," for a "large numbers of lives could be saved
by adequate fallout shelter space." The $3.5 billion venture would
cost $700 million in the first year, or a mere three to four dollars
for every American citizen. But even that cost was prohibitive.
Almost from the time the program was announced in January 1962,
the administration began to retreat. Continued reluctance on the
part of advisers—the President's science adviser among them—
slowed the effort and prevented implementation of the plan. Interest revived briefly at the time of the Cuban missile crisis (see
Chapter 7), when attack seemed closer than ever before, then

subsided as the crisis passed. Instead of a shelter program, the administration promoted less dramatic—and less expensive—initiatives. Exercise Spade Fork followed the Operation Alert activities of the 1950s in testing the state of readiness in a simulated nuclear strike. A booklet from the Department of Defense and the Office of Civil Defense entitled *Fallout Protection: What to Know and Do about Nuclear Attack* continued the Eisenhower administration's attempt to cajole the American public into taking action on its own.[39]

That booklet reflected the limitations of the administration's commitment. Its publication was clearly a way to honor the President's promise to help the public prepare for a nuclear attack, without providing further government support. It contained useful information about different kinds of shelters and ways to stock them and offered advice about finding fallout protection after a blast. But the whole effort was subdued, for as Adam Yarmolinsky, a Defense Department official, observed, "We wanted to be entirely honest with the people without scaring the pants off them." In the end, twenty-five million copies were printed for distribution through post offices and offices of civil defense. Government officials chose not to distribute the booklet to every household in the United States out of fear, according to Deputy Special Assistant for National Security Affairs Carl Kaysen, that it "would create very strong political demands for a Federal shelter building program on a large scale such as was never contemplated in the original Civil Defense message."[40]

Although the administration tempered its earlier commitment to an aggressive civil-defense program, the public remained aroused. Agitated by the initiatives of Kennedy's first year, Americans in the early 1960s were more interested in shelters than ever before. "At cocktail parties and P.T.A. meetings and family dinners, on buses and commuter trains and around office watercoolers," *Time* magazine noted, "talk turns to shelters." Advice columnist Ann Landers was a firm supporter of home shelters. AEC commissioner Willard Libby built his own shelter outside his West Los Angeles home. Nestled into a hillside and protected by railroad ties and bags of dirt, it cost a modest thirty dollars. For fifteen thousand dollars, Bernard Benson, a missile scientist and electronics manufacturer, built a shelter in Malibu that he stocked with food, water, and a 1925 edition of the *Encyclopedia Britan-*

nica—to help him start over again from an acceptable base. Thomas
J. Watson, Jr., chairman of IBM, offered the company's sixty
thousand employees loans of a thousand dollars each to build home
shelters and also sold them shelter supplies at cost. The Rabbinical
Council of America recommended that any Orthodox Jewish con-
gregation building a new synagogue include a fallout shelter and
urged that community shelters be built as well. Senator Russell
summed up widespread popular sentiment when he declared: "If
we have to start over again with another Adam and Eve, then I
want them to be Americans and not Russians, and I want them
on this continent and not in Europe."[41]

As in the 1950s, the commercial possibilities were overwhelming. In Dallas, the Acme Bomb and Fallout Shelters Company
expected to do a hundred thousand dollars' worth of business in
its first month. Atlas Bomb Shelter in Sacramento sold a 35-ton
prefab model for between five thousand and six thousand dollars.
"We haven't done any advertising yet," said Atlas head Frank
Ringer, "but even so, there's so much demand we can hardly keep
up with it." Once again, *Life* featured all sorts of models, with
plans detailing how shelters could be built. The OCDM provided
encouragement by listing recommended specifications, though not
all suppliers honored them or even knew what they were. Still,
widespread construction continued.[42]

Yet the shelters of the early 1960s would probably have been
no more effective than those of the decade before. When a brush
fire swept through Willard Libby's section of Los Angeles, his own
inexpensive sanctuary collapsed. The roof of a twenty-five-
hundred-dollar steel shelter built in Dallas caved in during a rain-
storm. Consumers Union, which had earlier helped spark concern
about strontium-90 in milk, now turned its attention to fallout
shelters. In a semi-serious report, it rated structures, just as it
rated all other products, according to cost, reliability, and predicted
performance and concluded that no shelter a consumer could afford
to build was "acceptable."[43]

While shelters had obvious limitations, Americans who built
them believed in them and became involved in caustic debates over
who should be allowed entrance. Some, like Mrs. Alf Heiberg,
former wife of General Douglas MacArthur, were extraordinarily
generous. She built a structure in Washington, D.C., in the 1950s
large enough to hold a hundred to a hundred fifty of her neighbors,

for, as she said, "I wouldn't enjoy sealing myself up if I knew my neighbors were being blown to bits." Others were more selfish. One Chicago suburbanite in the early 1960s summed up a spreading feeling when he declared: "When I get my shelter finished, I'm going to mount a machine gun at the hatch to keep the neighbors out if the bomb falls. I'm deadly serious about this. If the stupid American public will not do what they have to to save themselves, I'm not going to run the risk of not being able to use the shelter I've taken the trouble to provide to save my own family." Hardware dealer Charles Davis of Austin, Texas, stocked his shelter with four rifles and a .357 Magnum pistol, nestled behind a four-inch-thick wooden door. "This isn't to keep radiation out," he said, "it's to keep people out." Members of the clergy joined in the debates. The Reverend L. C. McHugh, a Jesuit priest, argued in an essay on "Ethics at the Shelter Door" that it was perfectly moral to defend private shelters at gunpoint and refused to give ground despite a flood of protest. Other clergymen saw the matter differently. Lutheran pastor John Simmons retorted: "Self-preservation is the first law of human nature. But there is a higher law—God's." Taken aback by the controversy his civil-defense proposals helped arouse, President Kennedy tried to temper the discussion. "Let us concentrate more on keeping enemy bombers and missiles away from our shores," he declared late in 1961, "and concentrate less on keeping neighbors away from our shelters."[44]

After hitting a high point in the early 1960s, interest in civil defense began to wane. The arguments of opponents who felt that protection was impossible became increasingly persuasive. SANE, outspoken about the dangers of fallout, argued eloquently in newspaper advertisements, pamphlets, and public rallies that "fallout shelters are pitifully inadequate protection against nuclear attack" and proclaimed that they tended "to obscure the unprecedented catastrophe that nuclear war would bring, and the efforts that must be made to avoid it." Only a peaceful world with internationally guaranteed disarmament could provide the security people sought.[45]

The same arguments could be heard elsewhere. "The fallout shelter program cannot protect—period," said Seymour Melman, professor of industrial engineering at Columbia University. Barry Commoner, professor of plant physiology at Washington University in St. Louis averred: "A nuclear war would be self-defeating. There is no way out through a shelter program." New Jersey governor Robert B. Meyner made the same point when he told a political gathering that "going underground is no answer." Even if people survived, he asked, "What kind of world would they come up to? What would they use for air? What would they use for food? . . . What would they use for people?" A 1961 *New York Times* poll of New Yorkers recorded a commonly held belief that "protection against a superbomb at this time is almost hopeless." A Queens accountant concluded that "survival with the fall-out . . . would be miserable. I'd rather be dead," while a Manhattan bank teller decided that he might "run under the bomb when it falls. Get it over with quickly."[46]

Some critics saw the issue in ethical terms. For them, serious questions of human purpose took precedence over a haphazard and ill-conceived effort to hide from the effects of nuclear war. "It is the morality of men and affairs which challenges us, not the morality of moles or other underground creatures, slithering in storm cellars," asserted Rabbi Maurice N. Eisendrath, president of the Union of American Hebrew Congregations.[47]

Others cited strategic arguments. Challenging civil-defense advocates who claimed that an enemy would be less likely to attack if it was clear that the nation could absorb a nuclear blow and still save its population, they retorted that civil defense only fueled the arms race. The Federation of American Scientists pointed out that "the existence of a shelter program might itself influence the likelihood of a general war." It could "lower the provocation threshold, which might of itself make war more likely." The Soviet Union, the scientists continued, might well consider that a shelter program was part of an American effort to strengthen its own position as a prelude to nuclear attack. Gerard Piel, publisher of *Scientific American*, argued along similar lines in a address to the Commonwealth Club in San Francisco:

> The civil defense program of our Federal Government, however else intended, must be regarded as a step in the escalation process. . . . It gives the sanction of action to the delusion that a thermonuclear war can be fought and survived. It encourages statesmen to take larger risks predicated on First-Strike Credibility and Post-Attack Recuperative Capacity.[48]

Meanwhile, technological improvements in weapons and delivery systems highlighted the criticisms. Bigger and better bombs—carried by bigger and better missiles—could reach the United States in minutes rather than hours, which meant that warning time was drastically reduced. In the face of such developments, doubts grew about the prospects of protecting the civilian population at all.

The Limited Test Ban Treaty (see Chapter 7), signed by the United States, the Soviet Union, and Great Britain in 1963, squelched the most serious agitation for civil defense in the postwar years. By ending atmospheric testing and eliminating the fallout which had aroused so much concern, it effectively neutralized proponents of protection and left civil defense a casualty of the arms-control process.[49]

In the years that followed, interest in civil defense waned. The public remained largely unconcerned once the fallout issue was resolved. The scientific community was preoccupied with other problems: weapons development, nuclear power generation, and the limitation of arms. Government officials, likewise, had other priorities—as they struggled with the Vietnam war and the resulting protest at home—that left them little time or energy for what many viewed as the frivolous programs of the past.

At periodic intervals after the Kennedy administration, American Presidents and their top aides asserted the importance of civil defense when it suited their larger strategic ends. Lyndon B. Johnson called "an effective civil defense program . . . an important element of our total defense effort." Gerald Ford declared that "our civil defense program continues to be an essential element of the nation's deterrent posture." But neither had any commitment to the kind of comprehensive program that had been proposed earlier.[50]

During the later 1960s, the Office of Civil Defense, located in the Defense Department, continued to survey, mark, and stock shelter spaces. While lobbying for a more extensive effort, officials in charge recognized that little more was possible as legislative interest faded. In fiscal year 1967, Congress rejected a ten-million-dollar request for an experimental program involving shelter subsidies. Secretary of Defense Robert S. McNamara, sympathetic to the notion of civil protection, settled for the limited program already under way. By the end of the decade, authorities had identified more than two hundred million spaces and distributed more than 165,000 tons of supplies.[51]

That approach continued into the early 1970s, as civil defense came to encompass peacetime disasters as well as those resulting from war. Melvin Laird, Secretary of Defense in Richard Nixon's administration, believed that with proper planning the nation could survive any crisis and argued that his department "can and should contribute to total civil disaster preparedness—civil defense and natural disaster." A Federal Preparedness Agency in the General Services Administration and a Federal Disaster Assistance Administration in the Department of Housing and Urban Development carried on activities related to those of the Defense Civil Preparedness Agency in the Department of Defense, until bureaucratic overlap led to demands for better coordination. In 1978, President Jimmy Carter authorized a stronger civil-defense effort, and in 1979, Congress pulled together all functions into a single organization, the Federal Emergency Management Agency (FEMA).[52]

Ronald Reagan gave civil defense new life. Committed to a massive defense buildup, the Republican administration that took office in 1981 was equally committed to a policy of surviving a nuclear war. Deputy Under Secretary of Defense T. K. Jones argued in an interview with *Los Angeles Times* reporter Robert Scheer that the American people could indeed withstand a nuclear attack. "Dig a hole, cover it with a couple of doors and then throw three feet of dirt on top," he said. "It's the dirt that does it. . . . If there are enough shovels to go around, everybody's going to make it." To that end, the administration decided to fund a major civil-defense program, despite the reasonable reservations of the Joint Chiefs of Staff and the Office of Management and Budget. In the 1983 fiscal year, it requested $252 million for civil defense, almost twice the amount of the preceding year. Meanwhile, FEMA began work on a seven-year plan estimated to cost $4.2 billion.[53]

The plan included a series of components. Assuming somewhat irrationally that warnings would provide a week's lead time prior to an attack, the administration revived the idea of evacuation and proposed moving millions of people into host areas in the countryside. "Sure, it'll be a hell of a mess," said FEMA official Louis O. Guiffrida. "You'll use anything at your disposal—trains, planes, cars, shoes. It'll be terrible. It boggles the mind. But do we just throw up our hands and say, 'Forget it, the job's too big'?

Do we give up?" Determined to take action, the administration also considered shelters once again. Facing a prohibitive estimate of seventy billion dollars for permanent shelters for the entire population, FEMA officials opted instead for temporary shelters in rural regions that evacuees themselves would help build. It welcomed the actions of private firms, like the American Telephone and Telegraph Company, that began to construct their own shelters to house top executives during an attack. At the same time, government officials drafted instructions for maintaining the economy after a war. A Federal Reserve System booklet asserted: "Victory in a nuclear war will belong to the country that recovers first." Federal Reserve banks would attempt to clear all checks, "including those drawn on destroyed banks," according to a National Plan for Emergency Preparedness. Credit cards would also be honored in a post-attack economy.[54]

Critics had a field day with the new plans. Columnist Ellen Goodman "couldn't decide whether to giggle or shiver" after reading through FEMA booklets. "The calm, chatty descriptions of how to survive nuclear war with just a touch of inconvenience had what Yale professor of psychiatry Robert J. Lifton calls 'the logic of madness,' " she wrote in one column. FEMA's theme song, she noted in another, was: "Pack up your troubles in your old family buggy and drive, drive, drive." Folksinger Fred Small took aim at T. K. Jones's naive approach to finding shelter as he advised listeners:

> Just dig a hole in the ground
> Climb right on down
> Lay some boards on top of you
> And sprinkle dirt around
> You won't have to be dead
> If you only plan ahead
> You'll be glad you kept a shovel on hand.

On a more serious note, Congressman Edward Markey told the House of Representatives that civil defense remained only "a band-aid over the holocaust," and other critics argued, as they had in previous decades, that civil defense could never provide adequate protection from the perils of nuclear war. Yet the government persisted in its approach for several years, until it became preoccupied, in Reagan's second term, with still another kind of protective shield.[55]

In the 1980s, as in the past forty years, civil defense was caught up in the larger debate over how to deal with nuclear arms. Scientists were active protagonists throughout the entire period as they considered the destructive consequences of burgeoning atomic arsenals. Their concerns helped spark the fears that developed in the 1950s and 1960s and the occasional pressure for personal and political response. That pressure, however, was never potent enough to prompt the massive civil-defense effort that proponents sought, despite serious interest that peaked at a number of points. Adequate protection of the civilian population in a nuclear attack remained an unfulfilled dream, a speculative and somewhat fanciful idea that never really caught on in the United States. National leaders, implicitly recognizing the myth that they could survive a cataclysmic war, proved reluctant to spend the huge sums required for protection that might prove meaningless in the end. They also hesitated to take action that would arouse even greater anxiety. "Are we to flee like haunted creatures from one defensive device to another, each more costly and humiliating than the one before," diplomat George Kennan asked on one occasion, and most of his fellow officials, Democrats and Republicans alike, answered no. And so defense planning retained a strictly military dimension— its focus on blunting or deterring an atomic attack. Protection would have to come from a broader effort to avoid nuclear war. Within that framework, civil defense remained something of an afterthought, useful as a signal of national resolve but hardly worth more effort or expense.[56]

6

The Peaceful Atom

Fear of nuclear catastrophe did not unfold in a vacuum. Hope for
a dazzling nuclear future offset the anxiety that accompanied news
of the first atomic bombs and increased with each passing year.
Nuclear power might counter the destructive possibilities of ever
more powerful nuclear weapons and provide an opportunity to
remake the world. So argued scientists and policymakers from 1945
on, as they came to rely on the new bombs that bolstered America's
defense but at the same time threatened an unforeseen calamity.
Most Americans shared their bifurcated views. Even as they contemplated the perils of fallout and debated the merits of civil defense, they considered the brighter side of splitting the atom. They
dreamed of a nuclear utopia, with electricity generated at virtually
no cost, with cars and planes and ships fueled by an inexhaustible
energy source, and with isotopes readily available for industrial
and medical use. For a time, however, the dream outstripped
reality. Scientists received greater rewards for research that contributed to the defense effort and, therefore, paid correspondingly
less attention to the technological problems that had to be overcome
to make the atom viable at home. Still the dream persisted, promoted by government officials who regarded peaceful applications

as the means to mitigate the potential horrors of the bomb. Throughout the late 1940s and 1950s, they speculated about the glorious age that lay ahead and encouraged similar speculation in the country at large. Children learned about the benefits of the peaceful atom at school, while the rest of the public encountered the same message in the popular press. The ubiquitous propaganda helped romanticize the atom and created an entirely new vocabulary for daily affairs. It also produced a sympathetic population pool as scientists and engineers made the technical breakthroughs that led to increasing orders for reactors in the 1960s and 1970s. But even as business boomed and popular support grew, a series of dramatic accidents in the 1970s and 1980s heightened concern about safety and shattered the dream of a peaceful nuclear world once and for all.

In the first year after the explosion at Hiroshima, many Americans set aside lurking fears as they contemplated the "golden age of abundance" that beckoned. Harry Truman waxed eloquent about the new possibilities: "God has given us through this tremendous discovery of atomic energy the unique opportunity to build one single human community, on the highest spiritual level, accompanied by unlimited material facilities." David Lilienthal reflected on the "almost limitless beneficial applications of atomic energy," while physicist George Gamow proclaimed: "The newly discovered possibility of liberating the hidden energy of uranium atoms promises us an almost unbelievable technical progress in the years to come." Gamow foresaw special fuel for transit, "a small package of which will be enough to fly a huge passenger airliner across the ocean. We may also prepare ourselves for a trip to the moon and to various planets of our solar system in a comfortable rocket-ship driven by atomic power." David Dietz, science editor for Scripps-Howard newspapers, provided an even more fanciful description of the new atomic age: "Instead of filling the gasoline tank of your automobile two or three times a week, you will travel for a year on a pellet of atomic energy the size of a vitamin pill. . . . Larger pellets will be used to turn the wheels of industry and when they do that they will turn the Era of Atomic Energy into the Age of Plenty." With its new atomic capability, the world could even control its weather, and "artificial suns will make it as easy to grow corn and potatoes indoors as on the farm." Others filled in further details of the vision. University of Chicago chancellor

Robert M. Hutchins predicted: "Heat will be so plentiful that it will even be used to melt snow as it falls. . . . A very few individuals working a few hours a day at very easy tasks in the central atomic power plant will provide all the heat, light, and power required by the community, and these utilities will be so cheap that their cost can hardly be reckoned." Power "too cheap to meter" became the dream of the day.[1]

In the years following World War II, AEC officials kept that vision squarely before the public. Gordon Dean, who succeeded David Lilienthal as chairman of the commission, cited the myriad benefits of atomic energy in speeches and articles in popular publications like *Parade* magazine. He spoke of "the Wisconsin farmer whose corn yields have been increased because agricultural research with radioisotopes has told him the best time and place to use his fertilizer each year" and of "the California oil company that developed a new engine lubricant with the aid of radioisotope research." He told readers about efforts to develop a nuclear-powered submarine and an atomic engine to drive military aircraft that would lead to more extensive civilian applications and underscored the theme that " 'The Atomic Age' is coming—to make your life better." Lewis Strauss, the next chairman, was an equally active propagandist as he professed his "faith in the atomic future" in an article in *The Reader's Digest*, just one of his many public efforts. "Our knowledge of the atom," he declared, "is intended by the Creator for the service and not the destruction of mankind." John A. McCone, who assumed the AEC chairmanship in 1958, followed suit with descriptions of "nuclear dynamite" useful in building harbors and canals and isotopes able to treat brain tumors and thyroid disorders and insisted on seeing the atom "as your protector and as your benefactor." Chemist Glenn Seaborg, who discovered plutonium and headed the AEC from 1961 to 1971, trumpeted his own vision of nuclear-powered plenty. He imagined a kind of "planetary engineering" that included cheap power for running homes and factories, energy for moving rivers and mountains, and "agro-industrial complexes" built around huge nuclear plants. He dreamed of shuttles to the Moon, atomic-powered artificial hearts, and plutonium heaters to warm the swimsuits of scuba divers. "Where science fiction goes, can the atom be far behind?" he asked. "My only fear is that I'm underestimating the potentialities."[2]

That same message filtered into public consciousness in still
other ways. In the summer of 1948, the AEC, the General Electric
and Westinghouse corporations, and the New York Committee on
Atomic Information sponsored a month-long exhibit on Man and
the Atom in Central Park. In the Westinghouse display, spectators
could watch a "chain reaction" with snapping mousetraps setting
one another off and witness a "real radiation detector" at work.
In the GE section, visitors received copies of *Dagwood Splits the
Atom*, a comic book that King Features produced after consultation
with the AEC. It showed Mandrake the Magician reducing Dagwood and Blondie to the size of molecules to show them the marvels
of the atom. When Mandrake sparked a chain reaction, Dagwood
behaved as always, rushing off in panic, yelling, "Blondie!" Hyatts-
ville, Maryland (as well as other cities) sponsored an Atomic Energy
Week, with displays that conveyed the same impression as the
Central Park exhibit. In January 1949, a special issue of the *Journal
of Educational Psychology* examined the assorted atomic-energy exhibits of the previous year. It cited General Leslie Groves's support
for such ventures and quoted him as saying that "the average
American . . . must learn that nuclear energy, like fire and electricity, can be a good and useful servant." That lesson was easy
to learn. As novelist Kurt Vonnegut, Jr., noted, through one character in *Player Piano*, published in 1952, "Atomic energy was
hogging the headlines, and everybody talked as though peacetime
uses of atomic energy were going to remake the world."[3]

Not all efforts to instill a vision of the wonders of the atomic
world were aimed at adults. Children in the 1950s enjoyed a number of breathless accounts that described glorious future possibilities. The dust jacket to *All About the Atom*, by Ira M. Freeman,
told young readers: "If you could magnify a single grain of table
salt until it became as big as the Empire State Building, each atom
in it would look only as big as the grain you started with! And if
you could place eighty million molecules in a row like marbles,
they would stretch only the distance of an inch." The entire book,
which described atomic fission, reactor design, and isotope use in
simple terms, made the same point as Glenn Seaborg in telling a
younger audience that "the things that happen in the fantastic
world of the atom sometimes sound like the wildest science fiction.
But they have been tested and are true beyond a doubt." In *The
Tenth Wonder: Atomic Energy*, Carlton Pearl took much the same

tack and told readers, "There were seven wonders of the ancient world. Everyone has his own favorite eighth wonder. Albert Einstein's theory of relativity is the ninth. And the development of atomic energy is the tenth."[4]

But the most popular account of all came from Walt Disney. The world-famous cartoonist—creator of Mickey Mouse, Donald Duck, and a host of other characters—produced an appealing film, *Our Friend the Atom*, that explored first the scientific background and then the architecture of the atom in terms that any child could understand. For those who may have missed the movie, Heinz Haber provided *The Walt Disney Story of Our Friend the Atom*, a book that was the most sophisticated and attractive of all the narratives directed at children. It began with a parable about an aged fisherman, who found a sealed vessel in the sea. When he pried the jar open, a long-imprisoned genie escaped, ready to kill his liberator. Granted one wish first, the fisherman asked the genie to prove that "one who is so mighty can indeed fit into such a vessel," whereupon the genie funneled back into the bottle. The fisherman quickly capped the jar but opened it once more when the genie promised to grant three wishes and "to make thee rich and happy all thy days." That fable provided a motif for the atomic age. "The story of the atom is like that tale," the book declared, and "we ourselves are like that fisherman." Scientists had opened a bottle when they split the atom and, in so doing, had released a terrible and threatening force. Now the people of the world, like the fisherman, stood, "marveling and afraid, staring at the results of our curiosity." Yet the narrative ended on an encouraging note: "The fable . . . has a happy ending; perhaps our story can, too. Like the Fisherman we must bestir our wits. We have the scientific knowledge to turn the Genie's might into peaceful and useful channels. He must at our beckoning grant three wishes for the good of man. The fulfillment of these wishes can and will reshape our future lives." The rest of the volume recorded the various ways the atomic genie might become a servant and friend.[5]

The classroom curriculum often took the same cheerful approach. The Atomic Energy Commission cooperated with the National Education Association in publishing *Operation Atomic Vision* in 1948. This ninety-five-page handbook for high school students sought to stress not the military prospects but the peaceful possibilities of atomic energy. "Why not keep the bright side of the atomic energy picture in the center of our attention?" it asked. Then it pictured the nuclear future in the same euphoric terms used by the most optimistic commentators: "You may live to drive a plastic car powered by an atomic engine and reside in a completely air-conditioned plastic house. Food will be cheap and abundant everywhere in the world. . . . No one will need to work long hours. There will be much leisure and a network of large recreational areas will cover the country, if not the world." Educational journals made the same point in 1948 and 1949. The teacher's emphasis, according to a University of Illinois education professor, should be "on the positive aspects of atomic energy control in the service of mankind."[6]

The public responded to the messages received from both public and private sources. An American Institute of Public Opinion poll in late 1948 revealed a strong belief that "in the long run, atomic energy will do more good than harm." Sixty-one percent of those with a college education held that proposition, while 18 percent did not; 47 percent of those with a high school education accepted it, while 25 percent did not; and only 31 percent of those with grade school training reacted positively, while 23 percent did not. (The rest of each group was undecided.) In 1950, the Survey Research Center at the University of Michigan reported that, despite some apathy, most Americans it had questioned believed they would be better off in the long run as a result of the discovery of nuclear power. Three years later, 68 percent of another national sample acknowledged that they had "seen or heard much about possible civilian uses for atomic energy—besides its military use." In a follow-up question, 60 percent had an awareness of its potential value in generating electricity.[7]

One reflection of public response was the ease with which atomic terminology became part of daily discourse. In the 1930s, the term "bombshell" began to be used casually to describe a sexy woman. It became an even more potent expression of female sexuality with the advent of the atomic bomb. Popular songs, likewise, gave atomic themes everyday currency. "Atomic Power," recorded by the Buchanan Brothers in 1946 (see Chapter 1), romanticized the new force in glowing terms and, like many other songs, endowed it with a divine dimension:

Oh, this world is at a tremble, with its strength and mighty power.
They're sending up to heaven, to get the brimstone fire.
Take warning, my dear brother, be careful how you plan.
You're working with the power of God's own holy hand.
Atomic power, atomic power, was given by the mighty hand of
        God.
Atomic power, atomic power, was given by the mighty hand of
        God.

Other songs in the next decade, like "Uranium," recorded by the Commodores, and "50 Megatons," recorded by Sonny Russell, kept the atomic vocabulary current. "Atom Bomb Baby," a rock & roll song recorded by the Five Stars in 1957, fused nuclear terminology and sexuality by finding a suggestive image for every atomic expression. The song began with the chorus that was then repeated after every verse:

Atom bomb baby, little atom bomb.
I want her in my wigwam.
She's just the way I want her to be.
A million tons hotter than TNT.

The verses themselves used even more direct nuclear images:

Atom bomb baby, loaded with power.
Radioactive as a TV tower.
A nuclear fission in her soul.
Loves with electronic control.

Atom bomb baby, boy she can start
One of those chain reactions in my heart.
A big explosion, big and loud,
Mushrooms me right up on a cloud.

Casual references helped justify the practical—and romantic—possibilities of atomic energy and made it an integral part of daily life. The new vocabulary encouraged an easy acceptance of the atom, which only had to be harnessed to make the world a better place.[8]

As Americans dreamed of exotic possibilities—atomic golf

balls that could always be found or nuclear-powered rockets that could fly deep into space—they tackled more practical problems as well. Project Plowshare, initiated by the AEC in 1957, aimed at using nuclear explosives to dam rivers, cut through mountains, and create canals. More promising was the effort to produce electricity with atomic power. The search for a new energy supply, begun soon after Hiroshima, stemmed not from an existing shortage but from a need to assert American leadership in still one more arena in the escalating Cold War. American science had created the first nuclear weapons; it should similarly be able to create the first nuclear plants to provide electricity. International prestige hung in the balance.[9]

The government assumed all responsibility for the effort. The Atomic Energy Act of 1946 that created the AEC gave the new agency full authority over military and civilian atomic development. It mandated the continued government monopoly over all nuclear materials, facilities, and experimental projects, and that policy effectively squelched private initiative. At the same time, the government's focus on military uses of atomic energy meant that it was less able to pursue peaceful applications aggressively; in the first years following World War II, it viewed the development of power as something of an afterthought, continued largely to provide the AEC with a softer image. Still, a foundation for a nuclear power industry was now in place.[10]

But progress came slowly. The AEC was overwhelmed with the task of managing the network of laboratories and manufacturing facilities it inherited from the Manhattan Project, especially as leading scientists left to pursue their own interests in civilian life. Those who remained were often preoccupied with military programs, particularly as a nuclear arms race began, and were correspondingly less interested in the prospects for power-producing reactors. When they did consider the generation of power, they painted a pessimistic picture of the obstacles that lay ahead.[11]

There were, according to AEC chairman David Lilienthal, "a whole mass of involved, difficult, scientific, technical, and industrial engineering problems" that had to be overcome in the first postwar years. Robert Oppenheimer, now head of the commission's General Advisory Committee, declared in 1947 that a nuclear reactor might provide usable atomic energy within five years and might use such energy within ten to twenty years to deal with "certain specific highly critical problems," such as generating power at the North Pole. But, he said, "I think that it will take between thirty and fifty years before atomic energy can in any substantial way supplement the general power resources of the world. That is under the assumption that development is pushed, that intelligent and resourceful people work on the job, that money is available for it." At the outset, it did not appear likely that Oppenheimer's conditions would be met. Physicist Eugene Wigner complained that there were few first-rate scientists in the early reactor effort. Enrico Fermi, who had built the first reactor at the University of Chicago during the war, considered the AEC's research program in this area overly bureaucratic and uninspired. "We despair of progress in the reactor program," Oppenheimer said in 1948.[12]

Despite wishful thinking about atomic power, only limited progress occurred in the Truman years. On December 20, 1951, the Experimental Breeder Reactor–1 (EBR–1) became the first atomic power plant ever to produce small amounts of electricity. Built by the Argonne National Laboratory at a site fifty miles west of Idaho Falls, the reactor generated enough electricity to power its own building—whenever it operated. While it proved that the creation of power was possible, it also underscored how much more remained to be done. Several similar programs produced other working reactors in the next few years, but none promised a significant supply of civilian power in the next decade.[13]

Prospects improved as Dwight Eisenhower assumed office in early 1953. Ike himself was intensely interested in atomic energy, both in its military and in its civilian applications. Eager for rapid development, he also displayed an unfeigned ambivalence, stemming from his awareness of the difficulties of harnessing the atom's extraordinary power. He hoped to use atomic energy to provide a more secure defense, even as he pushed for progress in disarmament talks, and he wanted to offset the creation of a stronger military capability with the development of a healthy nuclear power industry at home. Those aims called for a new approach to the entire problem. The Truman administration had failed to formulate a clear-cut policy to encourage civilian nuclear power, and its cuts in the reactor-development budget had made matters worse. Although private industry was showing interest in greater involvement, the AEC had proved indecisive in responding effectively to that interest. Ike was now determined to take the necessary steps to promote the atom's peaceful use.[14]

He announced his intended approach in a remarkable address at the United Nations on December 8, 1953. That speech had its origins in a proposal early in the year from a disarmament panel that included Oppenheimer, Vannevar Bush, and Allen W. Dulles (deputy director of the Central Intelligence Agency). The panel argued for greater understanding of the nature of nuclear escalation and recommended "a policy of candor toward the American people—and at least equally toward its own elected representatives and responsible officials—in presenting the meaning of the arms race." Much taken with the suggestion for candor, Eisenhower asked his speech writers for a speech that could provide such disclosure, but successive drafts over the next few months proved either too revealing or too dull. Meanwhile, in September, the President began to explore an idea for an even more positive approach that had occurred to him during a recent vacation. "Suppose," he asked an aide, "the United States and the Soviets were each to turn over to the United Nations, for peaceful uses, X kilograms of fissionable material." Lewis Strauss and Charles D. Jackson, the *Time* magazine editor who had joined the Eisenhower staff as a speech writer, took the suggestion and worked out the details of what came to be known as Project Wheaties. As one draft gave way to the next—and rewriting continued even as Eisenhower flew to New York to deliver the address—the proposal acquired far more substance.[15]

In the first part of his speech, Eisenhower reviewed the destructive potential of the nuclear arsenals of the world. "To pause there," he then declared, "would be to confirm the hopeless finality of a belief that two atomic colossi are doomed malevolently to eye each other indefinitely across a trembling world." Instead, he wanted "to help us move out of the dark chamber of horrors into the light, to find a way by which the minds of men, the hopes of men, the souls of men everywhere, can move forward toward peace and happiness and well-being." Eisenhower underscored his conviction that "this greatest of destructive forces can be developed into a great boon, for the benefit of all mankind. The United States knows that peaceful power from atomic energy is no dream of the future. That capability, already proved, is here—now—today."

He then proposed that the nuclear nations—the United States, the Soviet Union, and Great Britain—contribute uranium and other fissionable material to a new International Atomic Energy Agency under the direction of the United Nations. In this way, the world could begin to harness atomic power for peace. The speech, followed by thunderous applause, received an enthusiastic response from every corner of the globe.[16]

Eisenhower saw his address as a call for an "Atomic Marshall Plan" for the world. Just as the United States was helping fund and direct the reconstruction of Europe, so it would assist other nations to enter the nuclear realm by providing resources and expertise. In such a way, America could maintain its political and economic interests and ensure that an inevitable transition would take place on its own terms. The propaganda advantages were obvious. A National Security Council draft statement on policy a year later underscored how the program could "promote a peaceful world compatible with a free and dynamic American society." It could generate support for American purposes and counter the communists' claim that the United States was concerned only with nuclear destruction: "Atomic energy, which has become the foremost symbol of man's inventive capacities, can also become the symbol of a strong but peaceful and purposeful America." To that end, the United States participated in a successful technical conference in Switzerland in the summer of 1955 and in a power-reactor symposium in Belgium in the fall of 1956. The President made twenty thousand kilograms of U-235 available for distribution abroad, and the United States negotiated several dozen bilateral agreements for cooperation with nations around the world.[17]

At the same time, Eisenhower recognized that his foreign-policy initiative could help promote the growth of nuclear power at home. But any progress, in his view, required revision of the Atomic Energy Act of 1946. Its restrictive provisions needed to be loosened to enable private industry to plunge into domestic reactor development. In his State of the Union speech and budget message in January 1954, he gave notice of his intention to stress greater commercial involvement with peaceful programs for atomic energy. Lewis Strauss, an AEC commissioner from 1946 to 1950 and now Ike's nominee for chairman, followed the President's lead, for Eisenhower's policy dovetailed with his own passionate belief in the need to unleash "the genius and enterprise of American business."[18]

The administration moved quickly to make good on its promise and sent Congress legislation to amend the 1946 act. The Republican-controlled Joint Committee on Atomic Energy then drafted an entirely new bill and fought with Democrats over its provisions for the next six months. The Democrats argued that the proposal to invite private development of atomic power constituted a giveaway that would replace a government monopoly with a private one. After a lengthy filibuster, they were overruled, and Congress passed the Atomic Energy Act of 1954. The new statute made the AEC responsible both for promoting and for regulating the private development of nuclear power. It instructed the agency to issue licenses to private firms to build and operate commercial nuclear power plants and also directed the commission to adopt regulations "to protect the health and safety of the public." Despite that provision, the emphasis was on the promotional side. As Commissioner Willard Libby noted in early 1955: "Our great hazard is that this great benefit to mankind will be killed aborning by unnecessary regulation. There is not any doubt about the practicability of isotopes and atomic power in my mind. The question is whether we can get it there in our lifetime."[19]

A breakthrough in the development of peaceful atomic power came the same year that Congress revised the legislative guidelines. In this effort, administrative leadership was directly responsible for the desired scientific results. Navy captain Hyman G. Rickover had been working on the problem of nuclear propulsion since 1946, and his effort to build a nuclear-powered submarine laid the groundwork for the creation of reactor-generated power on land. Rickover, a career naval officer with a master's degree in electrical engineering from Columbia University, single-handedly oversaw the Navy's nuclear program. An ambitious, tireless administrator who refused to cut any corners or tolerate any sloppiness, he had personally directed the production of the electrical equipment American ships needed during World War II. In the process, he had developed close working relationships with private industry but had, at the same time, alienated fellow officers who rankled at his criticism of shoddy practices even outside his own domain and objected to his contempt for traditional military means of operation. Unworried about their disapproval, Rickover persisted in his proven approach to completing projects after the war as he became increasingly interested in the possibilities of nuclear power. In the postwar period, he accepted an assignment to the Manhattan Project facility in Oak Ridge, Tennessee, to learn what he could about atomic technology. He was convinced that a nuclear navy was possible and was determined to help build it.[20]

Rickover's passionate pursuit of his dream paid off. In Oak Ridge, he established the same kind of disciplined group he had formed during the war and paid the same meticulous attention to detail in examining various reactor projects. He also pressed his case tirelessly with both ranking officers and commission officials. Though neither the Navy nor the AEC at first appeared particularly interested in a nuclear submarine, his persistence established it as a priority. By 1948, Rickover held a dual role as head of the Navy's nuclear effort and branch chief in the AEC.[21]

Rickover was most interested in a thermal, water-cooled reactor, operating under high pressure, that could fit into the hull of a submarine. It would contain a small core of uranium fuel that would generate heat to boil water and create steam as it fissioned. The steam would drive a turbine connected to the submarine's propeller shaft. Insisting on the need to solve concrete engineering problems rather than engage in abstract theoretical research, he turned to industry for help and forged close ties with Westinghouse and General Electric, all the while keeping tight control over the entire operation. He and his staff reviewed every technical decision made and drove the contractors forward in an effort to have a functional nuclear submarine ready in five years. He met his goal with the successful launch of the Nautilus in 1954.[22]

The Navy program provided a workable model for civilian nuclear power. Even before the submarine's launch, the AEC settled on the pressurized water reactor as the centerpiece of a five-year research and development program. Although the agency had some qualms about building a full-scale plant that could not yet provide economical power, it was eager to move ahead. In September 1954, work began on the nation's first nuclear plant at Shippingport, Pennsylvania. The Duquesne Light Company offered to provide the site and to build and operate the generating facility. Westinghouse, which had played the largest role in producing the Nautilus, designed and constructed the plant's reactor.

The Navy's success in working with Westinghouse led the AEC to ask Rickover to assist the company in the Shippingport project.[23]

Once again Rickover came through, though this time he faced even more serious obstacles. Strikes, shortages, and jurisdictional disputes complicated already tight schedules, and his insistent manner continued to irritate those with whom he worked. According to one utility company executive, while "there is a certain grudging respect for Rickover's engineering knowledge and dedication to the job, he is generally regarded as such an egotistical SOB that progress has been made on these contracts despite his personality rather than because of it." Nonetheless, Rickover succeeded in bringing the plant on line at its full power rating just before the end of 1957. While the electricity it produced was expensive—sixty-four mills compared to six mills per kilowatt for conventional plants—its design and the data derived from its construction were enormously valuable for future projects.[24]

The major problem—over and above the cost of the power produced—was the fact that construction had been financed by the government and directed by a military officer. If the administration was to make good on its promise to promote atomic development by commercial industrial concerns, it would need to take a different approach. The public had to be prompted to accept the financial and developmental risks if nuclear power was ever to be more widely used. The Power Demonstration Reactor Program, announced at the end of 1954 just after work at Shippingport began, was an attempt to involve private industry more extensively in the production of nuclear-generated electricity. It was an effort, according to Senator Clinton Anderson, chairman of the Joint Committee on Atomic Energy, "to force-feed atomic development" with federal funds. The AEC would invite private companies to design and build nuclear plants that they would then own and operate. The agency would provide funding and assistance as needed and would also provide free fuel for seven years, with charges levied only for the material actually expended in the reactor. In the first year of the program, the AEC approved four proposals to build plants—one near Chicago, another near Detroit, a third in western Massachusetts, and a fourth in Nebraska. A second round of proposals brought the number of projects to seven. Despite that promising start, the AEC's efforts aroused little enthusiasm in Congress or the White House. Eisenhower had his eye on other more dramatic projects, while the National Security Council wanted to emphasize smaller reactors that might prove more attractive in foreign markets. It was clear that technology had not yet caught up with inflated public hopes and that, as the agency itself acknowledged, "competitive nuclear power is a number of years, perhaps 5–10, in the future."[25]

Impatient with the pace of progress, congressional Democrats who sat on the Joint Committee on Atomic Energy lobbied for a greater federal role. In the spring of 1956, a measure sponsored by Senator Albert Gore of Tennessee and Representative Chet Holifield of California sought to enlarge the AEC's reactor development program by directing the commission itself to construct six large-scale reactors around the country. Democrats questioned the commitment, and ability, of private enterprise to build a successful nuclear industry and worried that the United States would falter in a crucial Cold War contest. They charged that the administration's program consisted of "lofty words and little action" and left the United States "lagging in the world race for nuclear power, prestige, and world markets." After harsh debate, the Gore–Holifield bill passed the Senate but failed in the House of Representatives by a narrow margin. Having survived that challenge, the Eisenhower administration continued to push for private development.[26]

Two years later, as John McCone succeeded Lewis Strauss as AEC chairman, the agency faced still another challenge. The Joint Committee's *Proposed Expanded Civilian Nuclear Power Program* called for "positive direction" from the AEC in constructing twenty-one different reactors in the next five to seven years. In response, McCone initiated an internal policy review, which resulted in a ten-year plan, published in early 1960. It underscored the AEC's intention of making nuclear power competitive with fossil fuels over the next decade and outlined a three-part program that involved continuing experimental projects, building prototypes of the most promising reactors, and working with utilities and manufacturers to produce full-sized plants.[27]

The results were scarcely visible by the start of 1961, as the Kennedy administration assumed office. Only two reactors were operational: Shippingport, which was technically though not commercially successful; and Dresden, near Morris, Illinois, which had been shut down for several months in 1959 following defects that became visible during start-up tests. While twelve more small plants were under construction, federal funding was declining and electric-utility companies had no further building plans. "What has happened, most briefly, is that the glamour has gone out of atomic power," *Science* magazine noted in April 1962. The race for space had become the symbol of technological supremacy. "Accordingly, the goal of economically competitive nuclear power, once talked about almost in the way the race to the moon is discussed now, has lost much of its sense of urgency."[28]

The new administration needed to do something to promote continued nuclear development. Accordingly, the President asked the AEC along with the Department of the Interior and the Federal Power Commission to take "a new and hard look at the role of nuclear power." The AEC's assessment, *Civilian Nuclear Power . . . A Report to the President—1962*, argued that such power was "clearly in the short- and long-term national interest and should be vigorously pursued." Imminent breakthroughs promised to cut costs of future plants substantially. "We conclude that nuclear power is on the threshold of economic competitiveness," the report declared, "and can soon be made competitive in areas consuming a significant fraction of the nation's electrical energy." As energy needs grew, the document suggested, nuclear plants could offset the diminishing supplies of fossil fuels and by the year 2000 might provide as much as two-thirds of the nation's power. A federal program of subsidizing development should therefore continue, for "economic nuclear power is so near at hand that only a modest additional incentive is required to initiate its appreciable early use by the utilities."[29]

The administration proved unpersuaded by the report. National security advisor McGeorge Bundy argued that it failed to consider the larger picture of energy supply and demand, and President Kennedy declined to endorse its proposals. With the White House disinterested, the initiative for further development lay with the private sector alone.[30]

Private industry came through. In 1963, General Electric and Westinghouse launched a series of "turnkey" projects, in which they assumed all risks in building plants that utilities could simply take over. The Oyster Creek generating facility, built by General Electric for the Jersey Central Power and Light Company, was the first and best known of those plants. Jersey Central, which gambled that a nuclear plant could produce electricity more cheaply than a fossil unit, was to pay sixty-six million dollars, an estimated thirty million dollars less than it cost to build the reactor complex. General Electric was willing to let the plant serve as a "loss leader" to demonstrate the feasibility of nuclear power to other customers.[31]

The program was a stunning success. Public opinion became more sympathetic with increased favorable exposure. In early 1964, articles showing the positive side of nuclear power appeared in the *Atlantic Monthly* and *The Nation*, and the next year a similar piece in *The Reader's Digest* declared that the industry had "gone through its gawky, troubled youth and entered a promising young adulthood." A 1965 opinion poll, conducted by the Marsteller Research organization for the Babcock and Wilcox Company, which had built a reactor for a power plant in Buchanan, New York, showed that the public had become more accepting. In Buchanan, where a plant was already operational, 60.5 percent of the residents had positive impressions of nuclear power, compared to 3 percent who were negative. In Philadelphia, where no plant yet existed but public information programs had already paved the way, 60 percent were positive, against 4.5 percent who were negative. In Atlanta, where there was neither a plant nor a public relations program, 45.3 percent viewed nuclear power favorably, with 7.7 percent against it. The entire survey, *Nuclear News* reported, showed that "familiarity breeds confidence."[32]

Meanwhile, that sentiment translated into economic progress. In 1964, a number of utility companies began to examine bids from firms making reactors and the following year started to talk terms. Over the next few years, the AEC issued construction permits for several dozen plants, many of them much larger than previous ones, near the nation's major cities. In 1966 and 1967 alone, utilities ordered about fifty plants, which accounted for half the new power generating capacity planned during that time. As costs rose at the end of the decade, partly in response to the inflationary spiral spawned by the Vietnam war, the process slowed, but in the early 1970s, the nation's utilities began to place orders again, this time for still larger plants, and the oil embargo of 1973 promised even greater reliance on nuclear power. The peaceful atom had apparently come of age, as more than a hundred reactors were ordered in the United States between 1970 and 1974. Worldwide, the same pattern held; by the end of 1975, 157 plants were operating in nineteen countries, with even more reactors on order or under construction. Then the process slowed again and finally ground to a halt.[33]

Safety proved to be the sticking point. In the first years following passage of the Atomic Energy Act of 1946, safety was something of an afterthought. The AEC, most interested in atomic development, had no central regulatory office at all. When in 1947 it decided it needed to monitor reactor safety, it created a Reactor Safety Committee, an advisory group consisting of prominent atomic scientists and chaired by Edward Teller, to make recommendations to the agency's general manager. Though it performed an essential function, the unit often faced criticism for impeding growth. "The committee," Teller later recalled, "was about as popular—and also as necessary—as a traffic cop. Some of my friends, anxious for reactor progress, referred to the group as the Committee for Reactor Prevention, and I was kidded about being assigned to the AEC's Brake Department." Several years later, the AEC created another advisory group, the Industrial Committee on Reactor Location Problems, to deal with nontechnical issues pertaining to population, property, and geological prospects. In 1953, it merged the two groups into a single Advisory Committee on Reactor Safeguards.[34]

Safety became an issue of greater concern with the passage of the Atomic Energy Act of 1954. One important section of the new measure insisted on peaceful development "to the maximum extent consistent with the common defense and security and with the health and safety of the public." That mandate encouraged the AEC to create a unit to complement the part-time advisory group that continued to function. In 1955, it established a Reactor Hazard Evaluation Staff and later that year added a Division of Civilian Application to monitor safety implications of construction applications. The role of those sections was to provide "reasonable assurance" that a reactor could be run safely and that any unresolved safety problems could be handled during construction before a plant received an operating license. Regulation, with an eye toward safety, was now under way, but always the stress was on promotion rather than regulation of atomic energy. As Lewis Strauss told the Joint Committee on Atomic Energy the following year, AEC guidelines "should not impose unnecessary limitations or restrictions upon private participation in the development of the atom's civilian uses, . . . they should not interfere with management practices, and . . . such regulations should be enforceable in a practical and uniform manner."[35]

As a result, early standards were often flexible and imprecise. C. Rogers McCullough, head of the AEC's Advisory Committee on Reactor Safeguards, told the Joint Committee in 1956 that the "way we determine that the hazard is acceptably low is purely a matter of judgment." The conditional construction permits that the agency often issued contributed to the elasticity of the regulatory process. They provided a great deal of latitude in a pragmatic effort to ensure that a developer could proceed with the expectation that an operating license would be forthcoming in time. That policy signaled the AEC's willingness to do whatever it could to move ahead with construction as quickly as possible. While the agency recognized risks, it sought to minimize them. "Careful reactor design by competent people, careful, conscientious, and skillful operation, and adequate maintenance—and I would like to add, a good deal of luck," McCullough said, could prevent accidents.[36]

But even McCullough conceded that "there is some risk" of a mishap and acknowledged, "We cannot convince ourselves that it is zero." Scientists at the Brookhaven National Laboratory in 1957 examined the likelihood of an accident more systematically. Their report, known both by its AEC number—WASH–740—and by its title—"Theoretical Possibilities and Consequences of Major Accidents in Large Nuclear Plants"—was a worst-case study. It observed that while reactors would not explode like atomic bombs, smaller explosions might inadvertently release large amounts of energy. Such blasts might harm the reactor and workers at the facility and might also produce widespread radiation contamination in surrounding areas. Focusing on a hypothetical accident, the study examined several models to determine how much radioactivity might be released and how much damage might occur. It predicted that the worst possible accident at a typical reactor could cause thirty-four hundred deaths and forty-three thousand serious injuries. Property damage could range from half a million to seven billion dollars. To ease fears of financial ruin, the Price–Anderson Act of 1957 established that accident compensation would come from a reactor operator's private insurance, plus five hundred million dollars from the government, with neither the operator nor the government liable for further damages. Although the measure mandated a larger role for the public in licensing procedures, it was an explicit effort on the part of the government to remove impediments to atomic development as concerns about safety were becoming manifest.[37]

Controversy surfaced in various forums as scientists, politicians, and concerned citizens contemplated potential problems with nuclear power. One highly visible case unfolded in Michigan when the Power Reactor Development Company, a consortium headed by the Detroit Edison Company, sought to build a fast-breeder reactor named after Enrico Fermi in Monroe, near Detroit, in 1956. The breeder concept, first tested in Idaho several years earlier, relied on the fact that the fissioning nucleus of a uranium or plutonium atom emitted more than two neutrons. One neutron continued the chain reaction; the other neutrons could be captured by particles of fertile matter and so create new fissionable material to replace the depleted nucleus. In that way, a breeder could provide heat to drive power-generating turbines, all the while producing more new fissionable material than it used. The problem was that researchers were not persuaded of the safety of the breeder reactor. In a test at the Idaho experimental station in late November 1952, a power surge had caused a temperature rise that destroyed the core. Such problems needed to be resolved before building similar plants for widespread commercial use. Upon examining the proposal for the Enrico Fermi plant, the AEC's Advisory Committee on Reactor Safeguards recommended against moving ahead, but the AEC itself, committed to quick development, issued the permit nonetheless on the grounds that the company had offered "reasonable assurance that the reactor can eventually be operated safely." At that point, the United Automobile Workers, the American Federation of Labor, and the Congress of Industrial Organizations asked that the permit be suspended. Their efforts led to a legal battle that went all the way to the Supreme Court, which ruled in 1961 that the Fermi plant was acceptable as long as the AEC followed guidelines that had already been laid down and acted fairly within the framework of its own regulations.[38]

As new plant construction proceeded in the 1960s, nuclear advocates in both public and private sectors argued that nuclear power posed no appreciable risks, yet they failed to silence the critics. In 1965, Chauncy Starr, a leading industrialist, summed up the position of nuclear proponents when he declared before the American Nuclear Society, "Safety is a relative matter and I believe we have reached a point in the demonstrated safety of nuclear power to say nuclear power is safe, period." But not everyone agreed with Starr. Once concerns had been voiced—in the AEC, in other branches of government, and in the larger public arena—they could not be denied.[39]

Fears became more pronounced as accidents occurred at a number of plants. In January 1961, at Stationary Low-Power Reactor No. 1 in Idaho, the core went supercritical when a control rod was removed. The steam explosion that resulted killed two servicemen instantly, a third on the way to the hospital. All three bodies remained so radioactive that burial had to wait for three weeks. The troublesome Fermi plant suffered a similar problem five years later. In October 1966, as control rods were being withdrawn from the core to allow a nuclear reaction to occur, radiation levels suddenly rose precipitously. A small piece of zirconium had broken off, stopped the flow of coolant, and caused the partial melting of the core. Alarms went off, the process was halted, and the plant was shut down before further damage was done, but for a time there was concern that Detroit might need to be evacuated.[40]

In the mid-1960s, talk began about the "China Syndrome"— a sequence of events in which the mass of radioactive material in a reactor could overheat and begin to melt. In that scenario, the molten core, now out of control, would sear its way through the reactor floor, into the ground beneath, and on toward the general direction of China. Explosions might occur; at the very least, radioactivity would contaminate the surroundings. The larger the reactors, the greater were the meltdown fears.[41]

In response to mounting concern, particularly as plant size increased, the AEC issued a series of new regulations aimed at promoting safety. Previously, the agency had taken a case-by-case approach toward proposals for new plants; now it drafted broader guidelines dealing with emergency cooling systems, containment structures, and other required features that were designed to cover all contingencies. At first, the staff tried to avoid excessively restrictive terminology, keeping the agency's promotional function in clear view, but when tests at the not yet completed Oyster

Creek plant revealed production flaws, the AEC became more intent on ensuring quality control.[42]

At the same time, the government recognized the need to reassess safety risks. In March 1972, work began on a new study, directed by Norman C. Rasmussen, a nuclear physicist at the Massachusetts Institute of Technology. Although the AEC commissioned and paid for the survey, the agency did not orchestrate the three-year effort. About fifty outside consultants worked with ten AEC employees in the massive undertaking. Like WASH–740, this study concentrated on examining the probability of a nuclear accident and the consequences that might result. In the summer of 1974, Rasmussen released the first draft of WASH–1400, the *Reactor Safety Study: An Assessment of Accident Risks in U.S. Commercial Nuclear Plants.* The report was published in final form in October 1975.[43]

Several thousand pages long, the study found the risks of a reactor accident minimal. If a group of a hundred plants was considered, then the likelihood of a reactor mishap causing a thousand or more fatalities was "1 in 1,000,000 per year. Interestingly, this value coincides with the probability that a meteor would strike a U.S. population center and cause 1000 fatalities." In a worst-case scenario, the Rasmussen report said, thirty-three hundred people might die and another forty-five thousand might suffer from serious radiation-related illnesses.[44]

Sharp criticism both from inside and from outside the government appeared soon after publication of the draft. The American Physical Society issued its own report and pointed out that the Rasmussen study had neglected the cumulative death toll from cancer and leukemia caused by the release of radioactivity. The Environmental Protection Agency likewise found the estimates too low. A critique provided by the Union of Concerned Scientists and the Sierra Club argued that the chances of a major accident occurring were far greater than the report indicated. Although some of the figures were changed in the final version, the basic conclusions stood. The AEC was pleased with the report; other observers remained unconvinced.[45]

The critics were vindicated as accidents continued to plague nuclear plants. In March 1975, at the Brown's Ferry plant in northern Alabama, an electrician looking for air leaks with a candle inadvertently ignited some flammable insulation. The fire spread throughout the cable system and, after burning though the insulation, caused the cables to short-circuit. That damage eliminated controls for valves and pumps, wiped out instrumentation in the control room, and harmed the core cooling system. Eventually, operators managed to shut down both reactors, rig special pumps to provide water to cool the cores, and bring the blaze under control. But the fire still cost more than $240 million for repairs and replacement power and dramatized the ever-present risk of calamity.[46]

An even more dramatic accident occurred at the Three Mile Island plant near Harrisburg, Pennsylvania, in March 1979. Trouble began when a small amount of water leaked through a seal in the cooling system and led the pumps forcing coolant through the system to shut down. Core temperatures rose, and increased pressure opened up a relief valve and shut off the fission reaction. Unfortunately, other problems exacerbated what should have been a relatively minor accident. Two valves remained closed when they should have been open, and another failed to close as expected. Indicators on the control panel also malfunctioned, and operators compounded that failure by ignoring signs of abnormally high temperatures. Shutting down the emergency core cooling system in the erroneous belief that the core contained too much, rather than too little, coolant, they opened the way to a partial meltdown. Steam in the core combined with zircaloy in the fuel rods to cause first a small chemical explosion and then a large and potentially unstable hydrogen bubble in the containment building.[47]

The crisis had a major impact on local residents and on the American public as a whole. The disaster created an air of disbelief and led people to look to such imaginative accounts as On the Beach or A Canticle for Leibowitz as they tried to comprehend the situation. The film The China Syndrome, which had just opened in Harrisburg, provided another framework for fathoming the catastrophe. Starring Jane Fonda and Jack Lemmon, its focus on a fictional reactor crisis seemed to be an eerie coincidence. But area inhabitants were not content to sit back and simply watch the accident unfold; nearly a hundred fifty thousand fled their homes. Reporters flocked to the region and covered the accident as a lead story for the entire world day by day. Popular culture similarly seized upon the scenario in the months and years that followed. A Superman comic the next fall showed the hero flying over the

Three Mile Island towers. *Mad* magazine's Alfred E. Newman posed in front of the now infamous towers and said, "Yes, me worry!" A character in novelist Ann Tyler's *Dinner at the Homesick Restaurant*, set in Baltimore, downstream from Harrisburg, declares, "Nuclear accidents! Atom Bums! Just look at the facts; those folks in Hiroshima didn't get near as many side effects as expected! Want to know why? It was all that Japanese food with soy sauce. Plain old soy sauce. Keep a case of this around and you'll have no more worries over Three Mile Island." If nuclear plants were ever to enjoy public support, a more comprehensive regulatory system was necessary.[48]

After extensive investigation of the accident, it became clear that human error was as responsible as mechanical error for the breakdown. It also became evident that the basic nuclear safety-monitoring system itself was at fault. The Nuclear Regulatory Commission (NRC), formed in 1974 when the AEC split into two parts—the NRC to deal with regulation and the Energy Research and Development Administration to promote atomic progress—had never fully come to terms with safety questions that were becoming increasingly complex.[49]

Still another accident less than a decade later drew further attention to the risks of nuclear power. A huge Soviet reactor in Chernobyl, a small city near Kiev, suffered in April 1986 the worst nuclear disaster the world had yet known. A large loss of coolant in the reactor's core, perhaps caused by human error, led to a meltdown. Water remaining in the system turned to steam, reacted with graphite blocks shielding the reactor's uranium tubes, and produced highly volatile gases. A day after the incident began, the gases exploded, igniting the graphite, smashing open the reactor core, and demolishing the containment building. As the fire burned out of control, the uranium fuel, still fissioning, melted and released a cloud of radioactive smoke into the sky. Soviet officials cut off access to the plant and finally contained the fire, but not before it contaminated large areas of the world. Fallout, caught in prevailing winds, spread across Europe and around the globe. Some Americans masked their concern with black humor. "What's the weather report from Kiev?" one joke asked; answer: "Overcast and 10,000 degrees." "What has feathers and glows in the dark?" another asked; answer: "Chicken Kiev." But the jokes could not hide the seriousness of the situation. Europeans plowed up freshly

planted crops and banned imports of livestock and vegetables from
the east. Berkeley physicist John W. Gofman predicted that one
million people would develop cancer as a result of exposure to the
fallout, and five years later estimates of the cancer death toll ranged
from seventeen thousand to nearly half a million. "The central
lesson of Chernobyl," author Joyce Maynard noted just weeks after
the disaster, "is that accidents *can* occur." Public confidence, already shattered by Three Mile Island, fell even further after the
Soviet catastrophe.[50]

Meanwhile, a growing environmental movement provided another check on the development of nuclear power. Drawing on
lessons learned in the civil-rights movement, activists created an
awareness of ecological abuses that haunted the entire power industry. At first, most attention focused on fossil fuels, which had
a devastating effect on air quality. Nuclear power, by contrast,
appeared "safe, clean, quiet, and odorless," according to proponents. But then in the mid–1960s, a major debate erupted over
thermal pollution. Steam used to drive turbines producing electricity had to be cooled somehow and was usually deposited back
into the river or bay from which water was drawn. Fossil-fuel
plants, which functioned in a similar way, generated less heat and
released some of it through smokestacks. While the release from
nuclear plants was not radioactive, it nonetheless had an impact
on fish and plant life as it heated the surrounding water. In the
middle of the decade, the Fish and Wildlife Service urged the
Atomic Energy Commission to exert its regulatory authority over
thermal pollution, but the AEC claimed it lacked responsibility
for non-radioactive problems. The agency won a court battle that
upheld its claim of no jurisdiction but lost the public controversy
that followed. The issue became highly visible in 1969 with publication in *Sports Illustrated* of an article entitled "The Nukes Are
in Hot Water." Written by Robert H. Boyle, a senior editor and
avid conservationist, the piece was a caustic attack on the AEC.
"What literally may become the 'hottest' conservation fight in the
history of the U.S. has begun," Boyle wrote. "The fight is over
nuclear power plants and the damage they can inflict on the natural
environment. The opponents are the Atomic Energy Commission
and utilities versus aroused fishermen, sailors, swimmers, homeowners, and a growing number of scientists." Castigating the
agency for failing to deal with the problem, he noted charges that

Glenn Seaborg, the Nobel laureate who had discovered plutonium and now headed the AEC, had "yet to discover hot water." Though the article often resorted to caricature, it highlighted the issue of thermal pollution, particularly for members of Congress. Public pressure was clearly affecting the regulatory framework in important new ways. In 1970, both the Water Quality Improvement Act and the National Environmental Policy Act ensured that federal agencies would deal with the problem.[51]

Still another court fight was necessary to provide the final resolution. The National Environmental Policy Act required all federal agencies, including the AEC, to assess the impact of measures that might affect the environment. When the Baltimore Gas & Electric Company proposed building a nuclear plant at Calvert Cliffs, thirty miles from Washington on the Chesapeake Bay, concerned citizens asked that the problem of thermal pollution be considered. The AEC countered, as it had earlier, by saying it would not examine non-radioactive issues, whereupon the citizens took the AEC to court. In the *Calvert Cliffs* decision in 1971, Judge J. Skelly Wright of the U.S. Court of Appeals criticized the AEC's "crabbed interpretation" which, he said, made "a mockery" of the Environmental Policy Act. By the terms of his ruling, the commission would have to assess a broad range of environmental hazards. The decision placed additional pressure on an already overburdened regulatory staff and helped create a licensing backlog. Increasingly, environmental groups intervened in the licensing process and demanded changes. Expensive safety features, some of them poorly planned, were added to plants already under construction and inflated both the cost and the time required for plant completion.[52]

Several years after the *Calvert Cliffs* decision, a number of direct action groups challenged the spread of nuclear power even more aggressively. In the effort to preserve an ecologically balanced, egalitarian society, the Clamshell Alliance in New Hampshire provided the model for other protest organizations. Founded in 1976 after the Public Service Corporation announced plans for a nuclear power plant in the town of Seabrook, the initial group (named for the clams threatened by the proposed plant) included environmentalists, former antiwar activists, and members of the American Friends Service Committee. Committed to the principle of nonviolence, the Clamshell Alliance organized two small occupations, then arranged a mass occupation in the spring of 1977 that included about twenty-four thousand people, fourteen hundred of whom were arrested. Although the organization suffered an internal split a year later, it spawned a number of similar groups before it fragmented.[53]

The Abalone Alliance, established in 1976 in northern California, was the largest such coalition. It focused on preventing Pacific Gas and Electric (PG & E) from bringing the Diablo Canyon nuclear plant near San Luis Obispo on-line. Made up both of local residents and of ecologically oriented activists from outside the area, the Abalone Alliance held a number of occupations at the Diablo plant. Despite increased public awareness of the problems of nuclear power after the accident at Three Mile Island in 1979, the plant was licensed in 1981, whereupon the Abalone Alliance called for a massive occupation that led to nineteen hundred arrests. As that demonstration ended, a PG & E engineer announced that an important error in the blueprints required closing the plant indefinitely for repairs. Though Diablo ultimately went back on-line in 1984, the Abalone Alliance succeeded in its larger goal of souring public opinion on nuclear power.[54]

Disposal of nuclear waste became another point of contention at about the same time. Scientists and policymakers became increasingly concerned about the "back end" of the fuel cycle. Residual waste remained highly radioactive, in some cases for centuries, and needed to be discarded, but no one knew where. A report prepared in the spring of 1976 at the request of the Joint Committee on Atomic Energy described various options, none of them entirely satisfactory, while at the same time the new Energy Research and Development Administration recommended converting liquid waste into solid form, which could then be buried in a stable geological formation. But such a repository still had to be found, and the search for a satisfactory site generated furious controversy in the states under consideration.[55]

The various problems came to a head near the end of the 1970s. Although Richard Nixon had initiated "Project Independence" earlier in the decade to make the nation energy self-sufficient by relying heavily on nuclear power, now a combination of cost overruns and concerns about safety undermined the effort. As predictions of future energy needs were scaled down, new plants no longer seemed as necessary, and growing local opposition made their construction less likely. By the time Jimmy Carter became President in 1977, new orders for nuclear generating equipment in the United States had dried up. Soon, even plants under construction fell upon hard times.[56]

The 1980s proved to be dismal years for the nuclear power industry. In the summer of 1983, the Washington Public Power Supply System suffered a serious financial shock. Several reactors under construction in the state of Washington were declared superfluous and building was stopped. Without plants to generate electricity, there was no way to make payments on bonds for construction. The system defaulted in the largest municipal bond default in U.S. history. Other plants in the pipeline faced similar problems. In 1984, in Ohio, Cincinnati Gas & Electric announced that it planned to convert the 97-percent-finished William H. Zimmer nuclear plant into a coal-burning facility to cut down costs. That same year the nearly completed Midland plant in Michigan likewise faced conversion to fossil fuel. Licensing controversy became a way of life, and activists arguing that nuclear power was too dangerous to be tolerated on Earth engaged in adversarial hearings that sometimes prevented operating permits from being issued. A *Forbes* magazine cover story in 1985 called the nuclear industry "the largest managerial disaster in business history," while a *Washington Post* poll the next year found that in the wake of the Chernobyl disaster, 78 percent of the public opposed any new construction of nuclear plants. Despite Ronald Reagan's support for nuclear power as "one of the best potential sources of new electrical energy supplies in the coming decades," the dream of a cheap and inexhaustible power pool was effectively dead.[57]

Why had the dream failed? In part, because expectations for nuclear power had been overinflated from the start. A concerted campaign, orchestrated by eager government officials, had created public hopes for nuclear power that, in turn, generated pressures for progress that could not be sustained. Administrative expertise in the 1950s cut through scientific difficulties and brought the first reactors on-line. But the very spirit that propelled early development led AEC officials—and scientists working with them—to short-circuit the regulatory process in the hope that problems could be solved at a later date. In the 1960s and 1970s and 1980s, difficulties became increasingly visible and finally could no longer be ignored. Public fears, largely overshadowed by hopes for a glorious future in the first years of the atomic age, now became more focused. Those fears, once dormant, mounted in the face of ever more serious accidents and effectively undermined the push for further development. Meanwhile, images of various forms of catastrophe merged, and as concern about a nuclear holocaust increased, reactors often became surrogates for bombs. The nuclear genie, freed from the bottle in Walt Disney's fable, no longer held out the same promise. Americans now recognized, as Oak Ridge National Laboratory director Alvin Weinberg noted in 1972, that they had "made a Faustian bargain with society," and for many of them, that agreement now appeared less attractive than before.[58]

7

The Search
for Stability

The attempt to regulate atomic power paralleled the effort to avoid
a catastrophic nuclear war. As warheads and delivery systems
became more powerful in the 1960s and 1970s, the arms race that
scientists had predicted engendered costs that spiraled out of control. Policies that had been developed earlier against the backdrop
of the Cold War no longer seemed as compelling with far more
sophisticated—and widespread—weaponry. Although Dwight Eisenhower had authorized the doctrine of "Massive Retaliation"
against American enemies and acknowledged that the aim of military strategy was "to blow hell out of them in a hurry if they start
anything," even he had begun to question that approach as he left
office. The new Kennedy administration, still caught up in the
Cold War consensus, was nonetheless cognizant of the critique of
analysts who argued against such a policy of overkill, and it seized
the initiative in defining a more flexible military approach. At the
same time, it acknowledged the extensive public criticism of un-
restrained atomic testing and the poisonous fallout it produced and
responded with the first systematic attempt since 1946 to contain

nuclear arms. Yet even that effort went only so far. While the Limited Test Ban Treaty and the pacts that followed demonstrated that agreement among the great powers was possible, they were but the first steps in a larger campaign. For even though the nuclear threat may have begun to abate, the clock on the cover of the *Bulletin of the Atomic Scientists* still stood at seven minutes to midnight as the 1980s began.[1]

The American arsenal had become increasingly powerful in the 1950s. The hydrogen bomb had raised the stakes of nuclear destruction, and intercontinental ballistic missiles, like the one that launched *Sputnik I*, made the new weapons even more ominous. At the same time, the arsenal had grown far larger, with the number of warheads multiplying twenty-fold. Worst of all, the scientists engaged in the development process appeared preoccupied solely with advancing the nation's offensive strength. "Defensive weapons," physicist Freeman Dyson reflected in 1979, "do not spring like the hydrogen bomb from the brains of brilliant professors of physics. *Defense*," he went on, twisting Robert Oppenheimer's response to the thermonuclear breakthrough, "*is not technically sweet.*" As the arms race became institutionalized, *The New Yorker* noted later, "it seems as if it had become an alliance, a sort of perverse love match, a deranged crush, each side in passionate awe of the other's prowess, each side desperately trying to concoct new and even more intoxicating ways of holding the other's rapt, undivided attention."[2]

But the arms race was something more than a strategic minuet. It was a deadly serious contest for superiority with weapons that could obliterate the human race. Military and strategic analysts who became increasingly influential in the 1950s—members of the self-styled nuclear priesthood—were concerned about how the delicate balance between East and West might be maintained in the face of ever stronger weapons. They questioned the assumption that the policy of Massive Retaliation would promote international stability by deterring a nuclear attack and suggested less rigid alternatives instead. These analysts succeeded the scientists, like Oppenheimer and his associates, who had been most outspoken in the first decade of the atomic age. While the scientists had sought to check the spread of weapons of destruction, the new critics of American policy were concerned instead with refining the nation's strategic approach to make it more effective. As Bernard Brodie, Henry Kissinger, and Herman Kahn speculated about the possibilities of limited nuclear war, a number of other policymakers and theorists began to examine the limits of deterrence and to debate the appropriate course for the United States. Scientists still had a voice, to be sure, but other figures spoke more loudly now.

Paul Nitze was one of those critical of Eisenhower's defense policy. An investment banker before World War II, he had stood in the rubble of Hiroshima soon after the bomb fell and then coordinated the government's official Strategic Bombing Survey assessing the overall impact of the air war against Germany and Japan. He then went to work for the State Department, remained deeply concerned with the implications of nuclear weapons, and over the next several decades advised five Presidents about their use. Nitze was author of NSC–68, a National Security Council document presented to Truman in 1950 that argued for a dramatic effort to maintain economic and strategic superiority in order to counter the Soviet threat.[3]

Six years later and out of government for a time, Nitze challenged Ike's approach to nuclear weapons with an article in *Foreign Affairs*. In "Atoms, Strategy and Policy," published in January 1956, he charged that there was a contradiction between the government's "declaratory policy" and its "action policy." Though the nation was, in fact, interested in "graduated deterrence"—using only such force as was necessary to deter aggression—its rhetoric promised much more. Massive Retaliation—the declaratory policy—"suggested that we would no longer take the measures necessary to contain local aggression with graduated means but would choose unlimited city-to-city atomic retaliation the moment we were given an excuse." Such an approach was ineffective, for it deviated too far from the likely response. Atomic weapons, Nitze acknowledged in his essay, were here to stay and would affect all future strategy, whether or not they were used. "The situation," he suggested, "is analogous to a game of chess. The atomic queens may never be brought into play; they may never actually take one of the opponent's pieces. But the position of the atomic queens may still have a decisive bearing on which side can safely advance a limited-war bishop or even a cold-war pawn." The necessary key for Nizte was an action policy that articulated limited responses to Soviet threats and a declaratory policy more clearly in line with that course.[4]

Meanwhile, another expert pondered Eisenhower's policy from a different angle. Albert Wohlstetter, a logician trained in mathematics and philosophy who had worked as a quality-control expert with the War Production Board during World War II and had then moved to the RAND Corporation as a quantitative systems analyst, became concerned with what he considered America's strategic vulnerability and inadequate readiness for nuclear war. He made his views public in 1959 with an article, also in *Foreign Affairs*, entitled "The Delicate Balance of Terror," in which he argued that the Eisenhower administration's reliance on Massive Retaliation left it unable to respond adequately to the Soviet threat.[5]

Wohlstetter countered authorities who claimed that the international nuclear balance reduced the risk of war. He began by challenging the arguments of such officials as Winston Churchill in Britain, Raymond Aron in France, and George Kennan and Dean Acheson in the United States, who posited that mutual extinction would necessarily follow a nuclear clash. Their hope that deterrence based on mutual terror created a measure of stability, he argued, was unfounded, for the thermonuclear balance was not secure but was "in fact precarious, and this fact has critical implications for policy." It was difficult to analyze the choices that nuclear nations faced, and the attendant uncertainties undermined the equilibrium the great powers sought. Deterrence, in short, was "not automatic." Acknowledging that "deterrence is not dispensable," Wohlstetter nonetheless concluded that "we have talked too much of a strategic threat as a substitute for many things it cannot replace." It was essential to consider as well "a more serious development of power to meet limited aggression, especially with more advanced conventional weapons than those now available."[6]

A similar critique came from General Maxwell D. Taylor, Army Chief of Staff between 1955 and 1959. A career officer for thirty-seven years and the one who had led the 101st Airborne Division in the jump on Normandy during World War II, he had been a vocal advocate of military modernization upon assuming top command of the Army a decade later. Retirement from that position made him even more outspoken, as he called for a "complete reap-

praisal of our military strategic objectives in light of the changes
which have occurred in the world, and which have invalidated, in
my judgment, the dependence upon massive retaliation as the . . .
keystone of our strategic art." In his book *The Uncertain Trumpet*,
excerpted in *Look* magazine, he developed his argument further.
Eisenhower's New Look, he charged, "was little more than the
old air power dogma set forth in Madison Avenue trappings."
Despite its public acceptance—stemming from frustration at the
Korean war, eagerness for a balanced budget, and characteristic
American faith in simple solutions—neither he nor the Army had
ever been happy with the rigid approach. "In its heyday," Taylor
argued, "Massive Retaliation could offer our leaders only two
choices, the initiation of general nuclear war or compromise and
retreat." But local conflicts were increasingly likely and would
require conventional American forces to deal with far stronger
Soviet ground troops. Reviving suggestions he had made earlier,
he proposed "a National Military Program of Flexible Response"
which included better training and preparation of a wide range of
forces able to fight a limited war.[7]

   The new Kennedy administration that assumed power in 1961
accepted the outlines of that argument. Kennedy himself was eager
to revive a sense of national purpose and to take a more energetic
approach to defense. He also wanted options that gave him "a
wider choice than humiliation or all-out nuclear action." Members
of his team, "the best and the brightest" (in author David Halberstam's wry phrase), shared his enthusiasm for an activist approach and his readiness to consider alternatives to current
strategic policy.[8]

   McGeorge Bundy, Kennedy's Special Assistant for National
Security Affairs, was one of those ready for change. A Republican
who had twice voted for Eisenhower, he had nonetheless developed
a close relationship with Kennedy in the late 1950s that drew him
into the new administration. As the bright, articulate, occasionally
arrogant dean of Harvard College, he had helped reshape the university; he hoped to play the same role in the federal government.
Like others in the Defense Department and on the National Security Council, he argued that the New Look had limited the
effectiveness of American policy. "We believed that conventional
military strength had been badly neglected in the Eisenhower
years," he later recalled. He was particularly critical of the rigidity

of the Single Integrated Operational Plan (SIOP) his predecessors had developed, which offered no alternative to overwhelming destruction if the United States launched an attack. The current war plan, SIOP–62, he told Kennedy, was "dangerously rigid and, if continued without amendment, may leave you with very little choice as to how you face the moment of nuclear truth." The difficulty was that it "calls for shooting off everything we have in one shot, and it is so constructed as to make any more flexible course very difficult."[9]

Secretary of Defense Robert S. McNamara shared Bundy's concern. First a faculty member teaching accounting at the Harvard Business School, he had then served in the Air Force during World War II, during which time he had helped apply business methods to operational plans in moving men and matériel to the front. At the war's end, he had accepted a position with the Ford Motor Company, as one of the "Whiz Kids" hired to turn the ailing firm around. There the formidable McNamara had become something of a legend, paying meticulous attention to detail, increasing quality without increasing cost, and rising to become president of the company. At the Pentagon, he hoped to use similar techniques to reshape military policy. Though he had not seriously considered questions of nuclear strategy before—he had read but one book on the subject, Henry Kissinger's *Nuclear Weapons and Foreign Policy*; and but one article, Albert Wohlstetter's essay on "The Delicate Balance of Terror"—he quickly became aware of the dangers faced in planning for possible use of the bomb.[10]

Soon after assuming office, McNamara reported critically on the policy he found in place. The strategy of Massive Retaliation, he told the Cabinet and the President, was "believed by few of our friends and none of our enemies" and led to "serious weaknesses in our conventional forces." Systematic as always, he called for a full-scale review of American policy, to be completed within a month, as a prelude to drafting alternative plans. McNamara appointed Paul Nitze to head the effort and enlisted a number of RAND theorists to assist him, though some, like Wohlstetter, declined to come to the Pentagon. Responding to ninety-six questions asked by the Secretary of Defense, the experts worked with a sense of excitement in what one called "an intellectual crap game." For the first time, they had the chance to formulate real policies that they had only speculated about before. They responded

with the critique the administration sought and with recommendations for a less rigid approach to defense. By the summer of
1961, the administration revised the SIOP to give the President
five options rather than two in case of a Soviet attack.[11]

Instead of Massive Retaliation, the administration committed
itself to a policy of flexible response. Rejecting the "absolute"
definition of deterrence of the Eisenhower years, with the threat
of a full-scale nuclear counterattack to keep the peace, it endorsed
the concept of "graduated" deterrence, which implied a defense
force that could wage war with conventional as well as nuclear
weapons. If the United States lacked that versatility, McNamara
said, "the Soviets may well conclude that they can use lesser forms
of military and political aggression without the danger of an all-
out nuclear response." Such a non-nuclear buildup was necessary,
he went on, for "escalation to a higher level of war is probably
more likely to occur if we go into these limited actions 'ill equipped
and ill prepared' to support the political positions and political
objectives that have been previously established." To that end,
Kennedy delivered a series of messages in his first few months in
office calling for increased airlift capacity, recruitment of more
men for the Army and Marine Corps, expansion of the nation's
anti-guerrilla effort, and additional research on non-nuclear weapons. Maxwell Taylor played an influential role in the new administration. He became first a personal adviser to the President on
military matters, then Chairman of the Joint Chiefs of Staff,
and was finally able to implement the views he had recorded in
The Uncertain Trumpet. The bomb remained a mainstay of the
American arsenal, but now other weapons were part of the larger
mix.[12]

Plans for possible use of nuclear weapons unfolded within
this larger framework. For a time in the summer of 1961, the top
leaders—Kennedy, McNamara, Taylor, Bundy, and several others—toyed with a study suggesting that a first strike against Soviet
military forces was feasible. But Theodore Sorensen, White House
counsel and speech writer, was appalled; Paul Nitze voiced his
fierce opposition; and McNamara squelched the plan. In place of
a first-strike approach, the Secretary of Defense recommended a
more defensive response with selective—rather than massive—
use of the atomic alternative if other resources failed to parry an
attack. Still retaining the earlier targeting focus, McNamara pro-

claimed his position publicly in a commencement address at the University of Michigan in June 1962 when he said that the "principal military objectives, in the event of a nuclear war . . . should be the destruction of the enemy's military forces, not of his civilian population." Such "counterforce" targeting, however, required an adequate nuclear capability, and so McNamara authorized Minuteman and Polaris missiles in quantities similar to those mandated by the Eisenhower administration.[13]

Later McNamara shifted course. He faced mounting opposition to the costs of the conventional buildup and was similarly concerned about large-scale spending on strategic systems to counter growing Soviet nuclear capability. He and others in the administration also came to understand more fully the implications of an atomic war. Smaller tactical nuclear weapons would not guarantee a limited nuclear conflict, and there was no reliable way of ensuring that a nuclear struggle with the Soviet Union could be kept within reasonable bounds. In the mid-1960s, therefore, McNamara began to consider the alternative of "countervalue" targeting that moved away from the enemy's forces and guaranteed instead the destruction of much of the Soviet Union's industrial power and population in a retaliatory strike. No longer could the United States hope to survive simply by eliminating the enemy's strategic reserves; it now accepted the need to threaten a wider range of targets to deter an attack. Though the bomb was still but one item in the menu of options, it was clearly the backbone of the reformulated approach to deterrence. The nation's nuclear forces were intended to prevent the other side from using its might in what would be an unwinnable conflict for both. "Mutual assured destruction" (MAD) became the slogan of the day.[14]

That shift occurred in response to the foreign-policy crises the Kennedy administration faced. In the struggle over Berlin that heated up in 1961 (see Chapter 5), war threatened as Soviet and American tanks faced each other for sixteen hours at one checkpoint along the newly built Berlin Wall until the Russians backed off. In the end, Soviet leader Nikita Khrushchev, who had provoked the confrontation, softened his insistence on an immediate peace treaty with East Germany and let the crisis pass, but not before American leaders had time to consider what could have happened had nuclear weapons been employed. They concluded that the risks of using their atomic capability outweighed potential results. They were determined to stand firm, to resist being taunted by Soviet threats, yet were nonetheless unwilling to exploit their own nuclear superiority in ways that might produce a violent war.[15]

The administration came to the same conclusion during the Cuban missile crisis. In October 1962, American air photographs revealed that the Soviet Union was making good on a pledge to support Cuba after the ill-fated CIA-backed attack at the Bay of Pigs. Offensive missiles assembled in Cuba did not upset the strategic balance, for the Soviets could already damage American targets with missiles based farther away, but the siting of the new weapons just ninety miles from American shores served as a direct challenge to the United States. In office less than two years, Kennedy felt he had to respond. For thirteen days, the world watched as the superpowers stood "eyeball to eyeball" in what British Prime Minister Harold Macmillan called "the most dangerous issue which the world has had to face since the end of the Second World War."[16]

Throughout the crisis, the Executive Committee of the National Security Council debated the options. Some members wanted an air strike—with conventional weapons—to knock out the sites, but Attorney General Robert Kennedy rejected such a plan by recalling the Japanese attack on Pearl Harbor in 1941 and declaring that his brother would not be the Tojo of his day. Maxwell Taylor and Robert McNamara urged that preparations be made for an invasion, which the Secretary of Defense considered "almost inevitable." Virtually everyone veered away from the notion of a deliberate, controlled nuclear attack.[17]

The American advisers recognized that even the inadvertent use of nuclear arms might lead to unmanageable escalation. At one point during deliberations, former Secretary of State Dean Acheson joined policymakers and declared that the United States would have to knock out the Soviet missiles. When asked what would happen next he responded: "I know the Soviet Union well. I know what they are required to do in the light of their history and their posture around the world. I think they will knock out our missiles in Turkey."

"Well, then what do we do?" someone else inquired.

"I believe under our NATO treaty, with which I was associated," Acheson answered, "we would be required to respond by knocking out a missile base inside the Soviet Union."

"Then what do they do," was the next question.

"That's when we hope," Acheson said, "that cooler heads will prevail, and they'll stop and talk."[18]

Since no one was willing to play out that scenario, Kennedy followed another course. He announced a naval blockade around Cuba to prevent the arrival of any more missiles—though he called the move a quarantine, for a blockade was an act of war. At the same time, in a television address aimed at both the nation and the world, he made it clear that while he would not initiate an atomic attack, he would not hesitate to respond to a nuclear strike: "We will not prematurely or unnecessarily risk the costs of world-wide nuclear war in which even the fruits of victory would be ashes in our mouth—but neither will we shrink from that risk at any time it must be faced." To make sure there was no misunderstanding, he declared that "it shall be the policy of this nation to regard any nuclear missile launched from Cuba against any nation in the Western hemisphere as an attack on the United States, requiring a full retaliatory response upon the Soviet Union." Then he waited as Soviet ships steamed toward the blockade. At the last moment, Khrushchev called the ships back, even while work on the missile sites continued. Finally, he backed off altogether, and the episode came to an end.[19]

The Cuban missile crisis was the most terrifying confrontation of the entire Cold War. While the administration had shied away from plunging ahead with a nuclear attack, the world was nonetheless closer to nuclear war than it had ever been before. Kennedy emerged as a hero, and his party benefited in the midterm congressional elections only weeks later. Yet as immediate relief began to fade, some critics charged that the President had acted recklessly with his willingness to move all the way to the brink and beyond. He had avoided disaster, as Dean Acheson wryly observed, by "plain dumb luck" and by Khrushchev's fortuitous decision to withdraw, but other, more violent outcomes might well have been the result.[20]

The crisis had the unintended consequence of pushing the Soviet Union toward a massive missile-building program. Earlier American fears about a missile gap that favored the Soviets had proven unfounded, and the confrontation in Cuba demonstrated American nuclear superiority. In the aftermath, the Kremlin began to amass a much larger arsenal of intercontinental ballistic missiles, ending up with fourteen hundred, the same number McNamara had called for in a five-year defense plan. The Soviets also built their own nuclear-missile submarines, like the American Polaris vessels, and began to bury land-based systems in hardened underground silos like those housing American Minuteman missiles. A far more vigorous arms race was under way.[21]

As policymakers played out their options in the continuing Cold War, other Americans responded more vocally to the atomic threat. Scientists had been chastened by the Oppenheimer hearings of the previous decade and remained cautious about voicing their views as aggressively as before. This time, the most acute criticism came from a different source. The public was already concerned about the impact of fallout and anxious about the limitations of civil defense (see Chapters 4 and 5). Now members of the creative community, accustomed to giving their imagination free rein, began to explore the possibility of a holocaust. Some commentators, like poet Robert Lowell, offered sober warnings of impending disaster. Lowell wrote in "Fall 1961":

> All autumn, the chafe and jar
> of nuclear war;
> we have talked our extinction to death.
> I swim like a minnow
> behind my studio window.
>
> Our end drifts nearer,
> the moon lifts,
> radiant with terror.
> The state
> is a diver under a glass bell.

Others used satire to confront problems more indirectly. Once again, songwriter Tom Lehrer was the most biting critic of American policy and its implications. As both France and China developed an atomic capability in the 1960s, he sang about the problems of proliferation:

> First we got the bomb, and that was good,
> 'Cause we love peace and motherhood.

Then Russia got the bomb, but that's okay,
'Cause the balance of power's maintained that way.
    Who's next?

Even sharper was his personal scenario for the next war:

So long, Mom
I'm off to drop the bomb,
So don't wait up for me,
But while you swelter
Down there in your shelter,
You can see me
On your T.V.
While we're attacking frontally,
Watch Brinkally and Huntally,
Describing contrapuntally
The cities we have lost.
No need for you to miss a minute
    of the agonizing holocaust.[22]

Meanwhile, the worlds of fiction and film provided an even more accessible perspective on conflict that might erupt at any time. In 1962, Eugene Burdick and Harvey Wheeler published the novel *Fail-Safe*, which sold several million copies and became the basis for a feature film released two years later. Through a fictional scenario, the authors examined "a problem that is already upon us"—technology out of control. The story begins with a minor malfunction at Strategic Air Command Headquarters in Omaha, Nebraska, when a tiny condenser burns out. Unfortunately, that glitch sends American bombers flying toward Russia, past the fail-safe points from which they can be called back. The American President is forced to tell the Soviet Premier that a terrible accident has occurred but that there is nothing he can do either to destroy the planes or to get them to return. In order to avoid a devastating war that could escalate out of control, he must exchange one city for another in an even swap—and so New York City is paired with Moscow—to keep the international balance intact. The final scene shows a bomb falling on New York. "The accident may not occur in the way we describe it," the authors acknowledged in a preface, "but the laws of probability assure us that ultimately it will occur."[23]

In addition to probing the implications of increasingly complex technology, Burdick and Wheeler explored the erosion of human accountability in *Fail-Safe*. Near the end of the story, the President reflects to his Russian counterpart, "This crisis of ours . . . In a way it's no man's fault. No human being made any mistake, and there's no point in trying to place the blame on anyone." Then he continues sadly, "The disappearance of human responsibility is one of the most disturbing aspects of the whole thing. It's as if human beings had evaporated, and their places were taken by computers. And all day you and I have sat here, fighting, not each other, but rather this big rebellious computerized system, struggling to keep it from blowing up the world." For all of their efforts, there is not much they can do. [24]

The novel gained a good deal of attention; the film, which appeared after the Cuban missile crisis, received more modest acclaim. Acknowledging the ever-present possibility of mechanical malfunction, Americans were less eager by the time of the film's release to play out the sober scenario of how the world might end. They accepted the possibility of a holocaust, to be sure, but favored a more frivolous fictional approach.[25]

They were particularly taken by the black humor of Stanley Kubrick's brilliant film *Dr. Strangelove or: How I Learned to Stop Worrying and Love the Bomb*. Released the same year as *Fail-Safe*, it was a more vivid and absurd, even if more pessimistic, tale of a mad world become prisoner to its monstrous machines. The specter of total annihilation runs throughout the movie, for the Doomsday Machine can destroy life on earth, yet much of the action is outrageous, and the characters of General Turgidson, General Jack D. Ripper, President Muffley, Premier Kissof, and Dr. Strangelove himself give a wry, often ridiculous, touch to unfolding scenes. Events occur at Burpelson Air Force Base, where the commander has gone crazy; in the War Room, where officials are worried about a "doomsday gap" rather than a missile gap; and in the cockpit of an American bomber, where deranged crew members coax their wounded plane toward its fatal destination. It ends with the triumph of the Doomsday Machine—the ultimate but unsuccessful deterrent—as the camera pans across a series of mushroom clouds spreading through the sky and the sound track plays the Hawaiian song "Someday We'll Meet Again."[26]

Dr. Strangelove carried the logic of the nuclear age to its ludicrous extreme. It took contemporary military strategy—at one point the demented Dr. Strangelove cites arguments made in Herman Kahn's book *On Thermonuclear War*—and made it appear preposterous. At the same time, it mocked the notion of a diabolical superdeterrent as it casually recorded the triggering of a series of atomic bombs. Its morbid humor forced Americans to contemplate from an altogether different perspective the decisions their leaders were making. While they laughed at the irrational antics that appeared on screen, more often than not they left the theater subdued and disturbed.[27]

The ferment from the Cuban missile crisis—and from the extended public response—created pressure for the first agreement to control nuclear arms. Increasing international concern about radioactive fallout provided further support for a measure that might check the arms race at last. Efforts to limit testing in the Eisenhower years had resulted in a voluntary moratorium that lasted for several years but ended early in Kennedy's presidency. Now, as nuclear war appeared to be a real, not simply a theoretical, possibility, policymakers responded to that pressure by renewing efforts to reach an accord. The missile crisis provided Kennedy with the political strength that allowed him to proceed as long as he moved with moderation and similarly led Khrushchev to consider arrangements that placed the superpowers on the same level.[28]

Kennedy was eager for an agreement. He recognized the awesome power at his disposal and the responsibility it brought. "If you could think only of yourself," he once remarked, "it would be easy to say you'd press the button, and easy to press it, too." But the prospect of wholesale human slaughter following from such an action made it more intimidating. Pondering a strategy that accepted three hundred million deaths, he could only say, "And we call ourselves the human race." At the same time, he personalized the possibility of destruction with reference to his own family. "It may sound corny," he told diplomat Chester Bowles, "but I am thinking not so much of our world, but the world that Caroline will live in."[29]

The first steps toward an American–Soviet agreement came

as the missile crisis wound down. In one of his communications, Khrushchev wrote: "We should like to continue the exchange of views on the prohibition of atomic and thermonuclear weapons, on general disarmament and other problems relating to the relaxation of international tensions." In his response, Kennedy agreed that "perhaps now, as we step back from danger, we can make some real progress in this vital field. I think we should give priority to questions relating to the proliferation of nuclear weapons . . . and to the great effort for a nuclear test ban." In November 1962, the President again referred to the Cuban confrontation as he told Anastas Mikoyan, first deputy chairman of the Soviet Council of Ministers, "Look, this is an awfully dangerous world. I didn't think you would do this and you obviously didn't think I would react as I did. This is too dangerous a way for us to go on." In a speech to the Supreme Soviet the next month, Khrushchev agreed. The missile crisis, he declared, came close to being "like the tale in which two goats met on a small bridge over an abyss and, both of them refusing to make way, butted one another. As you know, both of them crashed into the abyss. Is it sensible for men to behave like that?" He wrote once more to Kennedy declaring that the "time has come now to put an end once and for all to nuclear tests, to draw a line through such tests." In a meeting with the director of the Arms Control and Disarmament Agency, an office created during his presidency, Kennedy underscored his own wish for a test-ban treaty.[30]

The sticking point was verification. The United States insisted on the need to conduct inspections on Russian soil to ensure compliance with any agreement, only to be countered by the Soviet Union's contention that such monitoring was little more than spying. In late 1962, Khrushchev gave ground and indicated his willingness to allow two or three on-site inspections, but Kennedy still held out for eight or ten. Even though scientists had provided a way out with the development of reconnaissance satellites that made it impossible for either nation to conduct tests without the other's knowledge, the impasse undermined the effort to reach an accord.[31]

In the spring of 1963, Kennedy decided to break the diplomatic deadlock. A short while before, he had asked Norman Cousins (Saturday Review editor and one of the founders of SANE) to sound out the Russian leader on the testing question while Cousins was on a trip to the Soviet Union. Cousins reported back that the Soviets seemed to be at a critical juncture and might well be nudged toward peaceful coexistence. Still sensitive to public pressure, the President chose the American University commencement in June to make a major address on the possibility of peace and so signal Khrushchev of his willingness to negotiate further.[32]

In his speech, Kennedy said that in the nuclear age "total war makes no sense." He spoke not of a Pax Americana but of a "genuine peace, the kind of peace that makes life on earth worth living, the kind that enables men and nations to grow and to hope and to build a better life for their children." He declared further that "world peace . . . does not require that each man love his neighbor—it requires only that they live together in mutual tolerance, . . . and history teaches us that enmities between nations, as between individuals, do not last forever." Noting the costs and consequences of the Cold War, he called for general and complete disarmament as an ultimate goal and asserted that a treaty outlawing nuclear tests was a necessary and possible step in that direction. To that end, he announced that the United States, the Soviet Union, and Great Britain had agreed to hold high-level discussions on such an agreement soon in Moscow and that the United States would not conduct further atmospheric tests as long as other nations refrained as well.[33]

While Kennedy's speech drew relatively little attention at home, it had the desired result on audiences abroad. England's Manchester *Guardian* called it "one of the great state papers of American history," and the Soviet Union proved equally responsive by publishing the address in its entirety and rebroadcasting the Voice of America translation. Khrushchev himself later termed it "the best speech by any American President since Roosevelt."[34]

Kennedy confirmed his commitment to negotiate by appointing W. Averell Harriman to work out the details in Moscow. The veteran diplomat, born into a wealthy railroad family, had long been more interested in power than money. During World War II, he had served as a special envoy for FDR, assuring Winston Churchill of America's commitment and support. Present at the creation of Cold War policies, he was not an idealogue but a man who recognized that shifting patterns required the ability to maneuver. After a term as governor of New York, he returned to Washington, where he worked as a roving ambassador and main-

tained the Soviet Union's confidence as he performed his assign-
ments. "As soon as I heard that Harriman was going," one Russian
diplomat told presidential assistant Arthur M. Schlesinger, Jr., "I
knew you were serious." Khrushchev agreed. "Harriman" he said,
"is a responsible man."[35]

Negotiations began with a sense of optimism. Kennedy himself, aide Theodore C. Sorensen later recalled, "was determined
to have a treaty, and arguments over language and wording, and
all the other dangers and disadvantages which might be pointed
out, could not deter him." If a comprehensive test-ban treaty was
still beyond reach, he was prepared to accept a limited agreement
instead.[36]

Within that framework, Harriman accomplished his task in
a matter of weeks. On August 5, Soviet, American, and British
negotiators signed the Limited Test Ban Treaty in Moscow. Although a preamble called for a broadly based disarmament pact,
the agreement itself was far more specific. It simply banned nuclear
tests in the atmosphere, or in any environment where detectable
radioactive debris might be spewed beyond territorial borders, and
sidestepped the inspection issue entirely. Portrayed as a step toward the goal envisioned by the Baruch plan in 1946, the treaty
was less an attempt to create a new international order than an
effort to deal with the special problems of the arms race. "I think
it is a very dangerous, untidy world," Kennedy said. "I think we
will have to live with it."[37]

Several days after the signing, the President sent the treaty
to the Senate for ratification. He was worried about its prospects,
for critics like Edward Teller were outspoken in their opposition.
The scientist who had played a major role in the development of
the hydrogen bomb remained obdurate in his quest for absolute
security, whatever the consequences, as he testified that the pact
was "a step away from safety and possibly . . . toward war." A
number of military officials and defense contractors, concerned in
part with their own prospects if the treaty passed, echoed his
argument. But Kennedy had made the necessary compromises by
assuring the Joint Chiefs of Staff that there would be no pause in
designing or testing new warheads, even if such tests now had to
occur underground, and he gained the support of Republican minority leader Everett Dirksen. The final vote was not even close;
the Senate ratified the treaty by a tally of eighty to nineteen. Most

scientists applauded and the clock of the *Bulletin of the Atomic Scientists* moved back five minutes to twelve minutes before midnight.[38]

Another agreement followed near the end of the decade. The Limited Test Ban Treaty had the effect of quieting public concern, particularly now that fallout would no longer sprinkle the globe. But policy analysts and their political superiors recognized that a host of problems remained. As proliferation became more and more of a problem, some of them spoke out. "With the kind of technology that is likely to be available in 1969," Herman Kahn declared, "it may literally turn out that a Hottentot . . . would be able to make bombs." Their concern prompted further efforts to address the problems of the arms race. Negotiating together after their first successful attempt to try to contain the spread of nuclear weapons, the United States and the Soviet Union finally reached agreement in 1968 and then invited other nations to join them in a Nonproliferation Treaty, which went into effect two years later. The three major powers agreed to resist aiding other nations from gaining nuclear arms, but were not obliged to renounce their own. By the time of implementation, ninety-seven nations had signed the pact and forty-seven had ratified it, though France simply agreed to abide by its terms without signing and China took no action at all. The *Bulletin* clock, which had edged forward to seven minutes before twelve as the nuclear club expanded, settled back to ten minutes before midnight after the ratification of the Limited Test Ban Treaty.[39]

Richard Nixon took the next step. Facing deep divisions in a country torn apart by the Vietnam war, the President who came to power in the turbulent electoral campaign of 1968 wanted to restore both domestic and international stability. Despite his lonely, almost humorless nature, which included a mean side that he tried to conceal, Nixon had an instinctive feel for political maneuvering. Like Kennedy, he was most interested in foreign affairs and was determined to leave his mark in this realm. The Red-baiter of the past, who had been an important part of the anticommunist crusade of the late 1940s and early 1950s, could afford to be flexible on Cold War issues, for his credentials were not subject to question. Suspicious at first of accommodation, he came to favor Strategic Arms Limitation Talks (SALT) as he gradually understood that a nuclear accord could rest at the center of a web of contacts with the Soviet Union. He and national security adviser Henry Kissinger, who had written astutely about nuclear issues in the past, saw arms control as part of the "linkage" tying all areas of foreign policy together. Recognizing that sufficiency, rather than superiority, was an adequate standard by which to measure American might, they became committed to a new course of détente and reversed the direction of American foreign policy followed since the end of World War II.[40]

The SALT I Treaty was a centerpiece of their policy. Though Gerard C. Smith, head of the Arms Control and Disarmament Agency, led the American delegation, the often sly and devious Kissinger retained a tight hold on the negotiating process. Just before the first session began, Nixon called in Paul Nitze, who had more experience in the area of arms control than anyone in the government and who then represented the Department of Defense in the talks with the Russians. Nixon confided that he did not trust Nitze's teammates and asked Nitze to report directly to him or to Kissinger. When difficult issues surfaced during the next couple of years of negotiation, Kissinger flew to Moscow to work through final details himself with top Soviet leaders.[41]

Negotiators for the Soviet Union and the United States tackled issues pertaining to both offensive and defensive capability. The United States had begun to consider a limited program of missile defense in 1967, and Nixon had supported a variant of this approach in 1969. Such programs, however, were controversial. Scientists questioned their effectiveness; opponents of military spending contended they were far too expensive; and inhabitants of areas considered prime locations feared they would become targets of attack. Furthermore, critics argued, an anti-ballistic missile (ABM) system might only encourage a nation confident of its own defense to launch a preemptive strike. Something needed to be done to limit such systems and to cap offensive programs at the same time, particularly as first the United States and then the Soviet Union embarked upon multiple independently targetable reentry vehicle (MIRV) technology that would allow a missile to contain several warheads that could go off in different directions in the course of an attack.[42]

After thirty months of negotiation in Helsinki and Vienna, a treaty was ready. Nixon flew to Moscow for a summit meeting with Leonid Brezhnev, First Secretary of the Soviet Communist party, and the two leaders signed the agreement; the U.S. Senate subsequently gave its overwhelming approval. The pact included an "interim agreement" on offensive forces, to last for five years, that set ceilings on intercontinental and other ballistic missiles in an effort to find a point at which the two nations were relatively evenly matched. More important was the ABM treaty that was part of the larger accord. It restricted each nation to the development and deployment of two anti-ballistic missile systems. In 1974 another agreement reduced the number to one.[43]

While the SALT I Treaty was a diplomatic success and a step toward disarmament, it hardly ended the threat of nuclear war. Both superpowers respected the limits imposed but moved to improve systems in other ways. "The way for us to use this freeze is for us to catch up," Kissinger declared soon after the 1972 summit, calling for qualitative refinements in American defense. "If we don't do this we don't deserve to be in office." Secretary of Defense Melvin Laird agreed and made Pentagon approval of SALT I contingent on a commitment to move ahead with a new B-1 bomber and a larger Trident submarine. MIRV technology developed rapidly. Though Kissinger acknowledged with uncharacteristic humility several years later that "I wish I had thought through the implications of a MIRVed world more thoughtfully in 1969 and in 1970 than I did," it was too late. The number of nuclear warheads in both American and Soviet arsenals more than doubled in the 1970s.[44]

Because the interim agreement was to last only five years, negotiations on a new pact began almost immediately. As scientists continued to improve the weapons strategists sought, diplomats tried to provide another check to the process. In 1977, President Jimmy Carter, echoing John Kennedy, offered his support to nuclear-arms limitation when he told the United Nations General Assembly that "nuclear war cannot be measured by the archaic standards of victory or defeat. This stark reality imposes on the United States and the Soviet Union an awesome and special responsibility." Talks culminated in the SALT II Treaty in 1979, signed again by Brezhnev on the Soviet side and by Carter for the United States. The new pact, far more complex than its predecessor, capped the number of warheads that could be placed on missiles, limited the number of multiple-warhead missiles, and froze the number of delivery systems permitted.[45]

While SALT II was another step forward, it met with a cool response in the United States. Hawks and doves alike were disappointed with its provisions. The Federation of American Scientists, normally supportive of initiatives to limit arms, opposed the treaty by claiming it only legitimatized arms competition. Hardliners argued that the pact gave too much away and was not verifiable. Carter's own party was divided, with people like Paul Nitze active in the Committee on the Present Danger, which called for strengthening, not limiting, American forces. Nitze claimed that the agreement simply ratified a Soviet buildup over the past decade. Opposition in the Senate was intense. By one count, only twelve votes were uncertain, and nine of those would be necessary for ratification. Carter did all he could and made initial gains, then watched support begin to erode. The Soviet invasion of Afghanistan in December 1979 killed any possibility of approval, and the President withdrew the treaty. Arms control would have to wait.[46]

The Reagan administration, which assumed power in early 1981, was unwilling to pursue the SALT approach. It struggled over whether or not to observe the limitations of the unratified treaty and waffled on the question. Its only real action was to rename the process by announcing that it would seek reduction, not limitation, of nuclear arms in Strategic Arms Reduction Talks (START). Yet there was little progress until Reagan's second term, when increasingly cool relations with the Soviets began to thaw and an Intermediate-Range Nuclear Forces (INF) Treaty eliminated one entire category of nuclear weapons.[47]

The halting effort to limit nuclear arms in the 1960s and 1970s reflected the larger struggle that began at the dawn of the atomic age. In the first years after Hiroshima, scientists had played a major role in mobilizing public opinion, as they tried to persuade policymakers that some kind of accommodation with the Soviet Union had to be made to head off a destructive arms race. They had failed then, as government leaders pursued their own less visionary course, and in the 1950s, after the Oppenheimer affair, the scientists had become less outspoken, though hardly less fearful of the consequences of a catastrophic nuclear war. While they assumed a lower profile, other Americans kept their concerns alive, and as technology provided bigger and better weapons in the decades that followed, new critics began to speak out. Some of those were strategists who argued for a more moderate military approach.

Others were members of the creative community who played an even greater role in reviving an awareness of the ever-present dangers of nuclear war. That concern finally propelled the arms-control process forward once more, though hardly as far as the critics wanted it to go. Bureaucracies were entrenched, diplomats remained rigid, and policymakers retained authority in their own hands. Still, critics had forced those in power to make a start. Later, when pressure could be brought to bear once more, they could take the next steps toward creating the stability that the world needed in order to survive.

A Resurgence
of Concern

The failure of the arms control process in 1979 prompted another wave of national concern. Several times in the 1950s and 1960s, the fear of atomic destruction had produced public calls for restraint. The fallout scare had led to a voluntary moratorium on atmospheric tests, while the Cuban missile crisis had sparked the first steps toward a nuclear agreement in the postwar years. SALT II, still another step in the arms control process, had offered further hope for continued progress in the effort to derail the drive toward nuclear war, until it ran afoul of both national and international politics. Its demise caused an instability that critics of defense policy felt compelled to address as the 1980s began. Scientists and other commentators alike were concerned with plans for ever larger arsenals that might be used at any time and with schemes that spoke optimistically about the prospect of limited nuclear war. "Planning on limited nuclear war is like planning to be a little bit pregnant," declared physician H. Jack Geiger, a member of one protest group. Once the process was under way, it moved forward with a logic of its own. A former Pentagon target analyst was equally tart in assessing the likelihood of top officials using all their weapons if a war broke out. "Look," he said, "you gotta understand that it's a pissing contest—you gotta expect them to use everything they've got." At the same time, critics questioned the leadership of the top policymakers in the United States. Ronald Reagan's relentless drive to bolster the defense establishment and his apparently cavalier acceptance of the possibility of nuclear war propelled them to speak out more aggressively than ever before. Their critique generated a groundswell of support as various groups bolstered one another in the move to revive the arms control effort.[1]

Although superpower talks had continued in the 1960s and 1970s (and the SALT I Treaty was the result), arms control activism had declined after the Limited Test Ban Treaty of 1963. Americans who had been worried about fallout and cataclysmic war seemed less concerned now that positive steps were being taken to ease the threat. Hiroshima and Nagasaki faded into the past as more attention focused on the Vietnam war. Then, toward the end of the 1970s, a new concern began to appear. Some people were worried about the safety of nuclear reactors. More were troubled by the failure of the arms control process, the buildup of atomic arsenals, and the apparent inclination of top officials to prevail in a nuclear war. Activists who had found protest dormant earlier seized upon the nuclear issue that now struck a resonant chord.[2]

They were particularly alarmed at Ronald Reagan's priorities in the realm of defense. Assuming power in early 1981, the new President was determined to eliminate "waste, fraud, and abuse" in domestic policy, even as he spent enormous sums of money for defense. With a militant approach toward the Soviet Union, which Reagan termed an "evil empire" in a casual reference to the popular *Star Wars* film, the administration sought an unprecedented one and a half trillion dollars over a five-year period to support a massive arms buildup. The nation was vulnerable, the President insisted, and so spending for weapons, both nuclear and conventional, had to increase. With his background in Hollywood and remarkable public relations skills, Reagan managed to deflect his critics and keep complaints over policy from diminishing his personal popularity.[3]

But opposition to his nuclear posture mounted. Six months before leaving office, Jimmy Carter had signed Presidential Directive–59 (PD–59), which called for increased targeting of Soviet military and political resources, improved the nuclear command and control structure, and committed the United States to fight a prolonged nuclear war that could last for months, not simply hours or days. Reagan took that policy one step further in National Security Decision Directive–13 (NSDD–13) in October 1981. Although the document remained secret, a classified five-year defense guidance that was leaked to the press provided the essentials. It stated that "should deterrence fail and strategic nuclear war with the U.S.S.R. occur, the United States must prevail and be able to force the Soviet Union to seek earliest termination of hostilities favorable to the United States." The nation, in short, now seemed committed to winning a nuclear war. That commitment, compounded by Reagan's career-long opposition to arms-control agreements and casual willingness to joke about pushing the button to launch a nuclear attack, helped mobilize opposition to his approach.[4]

In the early 1980s, scenarios of nuclear catastrophe became commonplace. Such speculative accounts of an atomic attack were nothing new, to be sure. Scientists and other authorities had long pondered the devastation likely in an atomic war. In 1948, the *Bulletin of the Atomic Scientists* published an article outlining the effects of a blast on a hypothetical American city, complete with a diagram showing degrees of damage in a series of zones. The next year, the Atomic Energy Commission considered the problems caused by an attack on Washington, D.C. Science-fiction writers had a field day with imaginative depictions of a world after nuclear war in stories and novels written after Hiroshima. In 1970, Hal Lindsey portrayed a cataclysmic nuclear conflict, "the final climactic battle of Armageddon," in *The Late Great Planet Earth*, which became one of the best-selling nonfiction books of all time. Now, however, far more numerous, realistic, and detailed descriptions of such a calamity began to appear. *Newsweek*, in 1981, published a "Scenario for a Limited War," complete with a box showing "Devastation over Detroit." Soon after, the *Eugene Register-Guard*, like many other newspapers, described a one-megaton bomb falling on the Oregon city's Center for the Performing Arts. In San Francisco, a booklet sponsored by the local government pictured what would happen if a similar warhead dropped on City Hall.[5]

One of the most extensive accounts of such catastrophe was

The Day After Midnight: The Effects of Nuclear War, based on a report by the U.S. Congress's Office of Technology Assessment. Published in 1982, it described "Possible Nuclear Wars" and had a lengthy section on "The Effects of Nuclear Weapons" that included diagrams showing the effects of a one-megaton surface burst, a one-megaton air burst, and a twenty-five-megaton air burst over Detroit. It printed estimates of both blast and radiation casualties and provided photographs of Japanese victims to remind readers of the human impact of the bomb. A final section on "Long-Term Effects" concluded that "the incalculable effects of damage to the Earth's ecological system might be on the same order of magnitude as the immediate effects of an attack."[6]

In that same year, the scientific journal *Ambio* pushed the descriptive process still further as it devoted an entire issue to the consequences of an atomic struggle. Articles dealt with a wide variety of topics: "The Biotic Effects of Ionizing Radiation," "The Impact on Ocean Ecosystems," "Effects on Agriculture," "Effects on Human Behavior," and "Epidemiology: The Future Is Sickness and Death." Yet the journal went beyond simple description alone; it broke new ground with its technical documentation and sophisticated analysis of potential catastrophe if nuclear weapons were used.[7]

Even more accessible were literary speculations about the impact of nuclear war. A growing number of artists and authors picked up on the arguments of worried scientists and so broadened the critique. By far the most compelling was Jonathon Schell's eloquent series of articles that ran first in *The New Yorker* and was then published as a book in 1982 as *The Fate of the Earth*. Schell, a staff writer for the magazine, drew on reports like "The Effects of Nuclear War" and books like *The Effects of Nuclear Weapons* to portray more vividly than anyone else the damage an atomic struggle could produce. His first chapter, "A Republic of Insects and Grass," was a graphic prediction of what the nation might become following a nuclear war. In terms a layman could understand, Schell began with the basic principles of radiation and summarized the immediate effects of a blast. He spoke as well of ancillary results and described the impact of fallout and electromagnetic pulse, which could wipe out all communication. He argued that the destruction of the ozone layer in the atmosphere could cause devastating climatic change. After describing the horrors of Hiroshima, he declared that "what happened at Hiroshima was less than a millionth part of a holocaust at present levels of world nuclear armament."[8]

Equally ominous was his second chapter, a philosophical speculation entitled "The Second Death." Grounded less in the cold, hard facts of physical destruction, this was an elaborate reflection about the extinction of the human race. "In a nuclear holocaust great enough to extinguish the species," Schell wrote, "every person on earth would die; but in addition to that, and distinct from it, is the fact that the unborn generations would be prevented from ever existing." Such a scenario was worse than death itself: "The thought of cutting off life's flow, of amputating this future, is so shocking, so alien to nature, and so contradictory to life's impulse that we can scarcely entertain it before turning away in revulsion and disbelief."[9]

The Fate of the Earth was not without its critics. John Leonard, a reviewer for the *New York Times*, argued that Schell simply "made the obvious less obscure" and offered no sense of how to deal with the dilemma he posed. Some scientists suggested that the atmospheric consequences might be different from those Schell described. Other readers found themselves moved by the description of destruction in the first chapter but less engaged by the contemplation of extinction that followed.[10]

In an effort to address his critics, Schell provided a sequel series of articles in *The New Yorker* in 1984, subsequently published as *The Abolition*. First, he returned "to the fantastic, horrifying, brutal, and absurd fact that we human beings have actually gone ahead and wired our planet for its and our destruction." Then he took up the question of "a deliberate policy" to abolish nuclear arms. The answer, he suggested, was "to preserve the political stalemate—to freeze the status quo" and at the same time to eliminate nuclear weapons, with the understanding on the part of the superpowers that such a step was really to everyone's advantage. For all of his noble rhetoric, Schell's effort fell short. The solution remained as elusive as it had been in his first book. As Lord Zuckerman noted in the *New York Review of Books*, "For those who might have been led to expect some shattering revelation," this was "a sad letdown."[11]

Criticisms notwithstanding, Schell's work was enormously popular. Like John Hersey's *Hiroshima* more than twenty-five years before, *The Fate of the Earth* was featured by the Book-of-the-Month Club and excerpted in the popular media. It became the focus of church sermons and community meetings around the country and helped create a diffuse but still real public sense that something needed to be done. While little that Schell said was new, his writing was riveting and mobilized the largest constituency yet to try to deal with the problems of the atomic age.

Meanwhile, numerous other authors offered fictional representations of the world after nuclear war. Such accounts were part of a long tradition of atomic-age literature in the post-Hiroshima years. Pat Frank, in *Alas, Babylon*, and Walter M. Miller, Jr., in *A Canticle for Leibowitz* (both published in the late 1950s), had explored questions of destruction and revival in books that gained a large following (see Chapter 5). In subsequent years, other writers followed suit. In 1964, in *Farnham's Freehold*, Robert A. Heinlein told the story of a family caught in a time warp after a nuclear blast. *Z for Zachariah*, published in 1975 by Newberry Award–winning novelist Robert C. O'Brien, described two atomic-war survivors trying to come to terms with one another in a world that would never again be the same.[12]

Such accounts became even more popular in the 1980s. Early in the decade, Pulitzer Prize–winner Bernard Malamud turned his attention to life after an atomic holocaust in *God's Grace*. At the start of his most original—and strangest—novel, he wrote: "At the end, after the thermonuclear war between the Djanks and Druzhkies, in consequence of which they had destroyed themselves, and, madly, all other inhabitants of the earth, God spoke through a glowing crack in a bulbous black cloud to Calvin Cohn, the paleologist, who of all men had miraculously survived in a battered oceanography vessel with sails, as the swollen seas tilted this way and that." Cohn, deep beneath the ocean during the catastrophe, struggles to re-create some semblance of society with a talking chimpanzee and other animals when he surfaces. Seeking to perpetuate at least a variant of the human race, he mates with a female chimp that he calls Mary Madelyn, only to have the other animals turn on him and his offspring, as the full cycle of savagery cannot be contained. Like *A Canticle for Leibowitz*, *God's Grace* was a depressing parable of a world that must bear the savage consequences of its inability to control its own self-destructive drives.[13]

Other novels explored related themes in these same years. Whitley Strieber and James W. Kunetka's *Warday*, a Book-of-the-Month-Club main selection, pictured in vivid detail the effects of a limited nuclear struggle. William Prochnau's *Trinity's Child*, a recommended choice of the Book-of-the-Month Club, showed the world in the midst of nuclear war. Like *Fail-Safe* in the 1960s, it described the military process moving out of control. Though Prochnau stopped short of the final cataclysm that other authors envisioned, he still conveyed a vivid picture of a world gone mad. For children, and their parents, Dr. Seuss, author of such classics as *The Cat and the Hat* and *Green Eggs and Ham*, published *The Butter Battle Book*. It told the story of an escalating arms race between the Yooks, who ate bread with butter side up, and the Zooks, who ate it with butter side down. As both groups created progressively stronger weapons, like the Triple-Sling Jigger, the Utterly Sputter, and the Bitsy Big Boy Boomeroo, the two sides faced off against one another and waited to see who would drop the bomb first.[14]

Popular musicians encouraged similar speculation about the prospects for human survival. Tom Lehrer enjoyed a revival of his iconoclastic verses in the 1980s, even as rock artists occasionally turned away from songs of love and sex and addressed possible cataclysm in their own way. In 1983, in the album "the final cut," Pink Floyd sang about "two suns in the sunset":

> in my rear view mirror the sun is going down
> sinking behind bridges in the road
> and i think of all the good things
> that we have left undone
> and i suffer premonitions
> confirm suspicions
> of the holocaust to come
>
> the wire that holds the cork
> that keeps the anger in
> gives way
> and suddenly it's day again
> the sun is in the east
> even though the day is done

> two suns in the sunset
> hmmmmmmmmmm
> could be the human race is run

Less lyrical than Lehrer, Pink Floyd reached a far larger audience and reflected the depth of public concern.[15]

Artists, too, shared their anxiety about the nuclear threat. In 1980, Alex Grey painted "Nuclear Crucifixion," which showed Jesus crucified in a mushroom cloud and conveyed a haunting message about the means of death. The following year, Robert Morris created a huge work called "Jornado del Muerto" (or "Journey of Death") after the New Mexico site of the Trinity test. Housed in the Hirshhorn Museum in Washington, D.C., it included a drawing of a Hiroshima bridge and photographs of Oppenheimer and Einstein, juxtaposed with a photo of a badly burned boy. Far simpler was Erika Rothenberg's acrylic done the next year called "Pushing the Right Buttons." It pictured nothing but two circular buttons, the top one labeled "Launch," the bottom one "Lunch," and dramatized fears caused by casual conversation at the top levels of government about the possibility of nuclear war.[16]

Filmmakers were equally active in playing upon nuclear themes. *Testament*, a television film which appeared in the fall of 1983, told the story of suburban housewife Carol Wetherly caring for her three children in the aftermath of a nuclear war that has killed her husband in nearby San Francisco. Based on a short story in *Ms.* magazine, the movie was a sober account of ordinary people responding bravely to chaos and certain death. It showed them behaving with dignity and courage in the face of catastrophe and provided a powerful affirmation of human life and love. Not long thereafter, the BBC in Great Britain produced *Threads*, another atomic-war film subsequently shown in the United States. Set in Sheffield, England, it described a Middle Eastern crisis erupting into a limited nuclear war that devastates England. Ruth, the movie's protagonist, has to function in a society in which supplies dwindle, population declines, and order breaks down. She gives birth to a child rendered retarded by exposure to radiation, and, as the film closes thirteen years later, her daughter bears a mutant child. Though similar to *Testament*, *Threads* was a more graphic, if less moving, account of the consequences of nuclear war.[17]

The film that received the greatest play, however, was the

1983 ABC television special *The Day After*. Focusing on the residents of metropolitan Kansas City, it showed missiles flying out of silos, bombs exploding, and bodies vaporizing on screen. Like *Threads*, it dramatized disruption more widespread than most people wanted to believe. Thanks to an intensive promotional campaign, the movie reached an audience larger than any of the other stories, novels, or films. It became an event even before it was aired, as *Newsweek* ran a cover story about it and analysts in communities around the country considered its potential effects. When the program was finally seen in the fall, ABC estimated that a hundred million people tuned in. In addition to a large audience, the film attracted its share of critics. Some complained that the movie slighted the political framework that led to war; others argued that survivors would never have been as alert as they appeared. For still others, after all the hype, the movie was something of an anticlimax. They were correct, of course; yet it still forced consideration of just what horrors the future might bring, and, as a commentator in *Time* magazine noted later, it reflected how the popular imagination, which had hesitated to confront the bomb in this way before, "seems desperate to embrace the thing today."[18]

All of these sources—the scenarios, songs, novels, and films—created widespread sentiment for doing something about the nuclear threat. As a result of the attention possible atomic destruction received, people were forced to confront the issue more directly. "I personally cannot walk down a street or look at a beautiful countryside or a painting or listen to a piece of music without being aware that this is possibly going to be destroyed," actress Joanne Woodward observed. For her, and for countless others, "It's either become active or crawl into a hole." Responding to the provocations—and to the protests—a number of different organizations launched drives to translate opposition feeling into meaningful change. Some of the activists were scientists; others were physicians; still others were ordinary citizens who felt a powerful need to do their part to ease the threat. They joined groups that cross-fertilized one another, participated in one another's programs, and together generated still further sentiment for aggressive government steps to ease the atomic threat.[19]

The first such group was Physicians for Social Responsibility (PSR). The organization had been founded in 1961 by Bernard

Lown, an articulate and intense Boston cardiologist and professor at Harvard's School of Public Health. A pioneer in coronary-care technology, he had become concerned about the arms race in 1959 when he heard a lecture about the possibility of a nuclear holocaust. He first organized a group of similarly concerned physicians in his living room, then arranged for publication of a series of papers in the *New England Journal of Medicine* in 1962 on the medical consequences of a nuclear war. That work underscored the inadequacy of any medical response to an atomic attack and helped persuade the Kennedy administration of the futility of bomb shelters. With the signing of the Limited Test Ban Treaty of 1963, PSR went into hibernation since the immediate threat seemed to have receded. Then in late 1979, as the arms-control process derailed and Ronald Reagan's campaign rhetoric revived old fears, Lown began to plan ways that physicians might take over from the physicists in trying to mobilize the public. "After all," he said, "if you have a serious problem, where do you go? In a secular age, the doctor has become priest, rabbi, counselor. Then, too, the doctor brings all the credentials of a scientist." As a first step, PSR persuaded the medical schools at Harvard and Tufts to sponsor a two-day symposium on "The Medical Consequences of Nuclear Weapons and Nuclear War."[20]

Held in February 1980, the meeting was a carefully orchestrated effort to gain public attention. It brought together some of the nation's leading arms-control advocates to address an overflow audience of six hundred—and through the media, the rest of the country. Organizers felt that it was essential to promote discussion of post-holocaust circumstances that most Americans seemed all too willing to ignore. As Howard Hiatt, dean of Harvard's School of Public Health, noted in his introductory comments, "What purpose, I wondered initially, to describe such almost unthinkable conditions? But the conditions are not unthinkable—rather they are infrequently thought about." Eric Chivian, staff psychiatrist at the Massachusetts Institute of Technology and PSR member who conceived and arranged the conference, underscored the need for more open communication and mobilization of public opinion. "The real problem is lack of information and misinformation at all levels," he said. "Many people believe nuclear war is going to happen and are so terrified they don't talk about it. But everyone wants to do something. The man in the street doesn't want his kids blown up. And that sentiment can be mobilized."[21]

The prevailing mood at the symposium was pessimism. Speaker after speaker addressed the devastation that would result from a nuclear war. One principal participant, George Kistia-kowsky, shared fears that stemmed from a lifetime of experience. Former head of the Manhattan Project's explosives division, later scientific adviser to Presidents Eisenhower, Kennedy, and Johnson, and finally a professor of chemistry emeritus at Harvard, he reluctantly assumed that atomic bombs would one day be used: "I think that with the kind of political leaders we have in the world . . . nuclear weapons will proliferate. . . . I personally think that the likelihood for an initial use of nuclear warheads is really quite great between now and the end of this century, which is only twenty years hence. My own estimate, since I am almost eighty years old, [is that] I will probably die from some other cause. But looking around at all these young people, I am sorry to say that I think a lot of you may die from nuclear war." Other participants described the devastating effects of modern weaponry, outlined scenarios of catastrophe, and concluded that medicine could do little to alleviate the likely chaos and suffering.[22]

The PSR meeting did a good job in drawing attention to the perils of nuclear war. Its organizers' basic position was that only such awareness could lead concerned citizens to press their governments to shift direction and thereby minimize the threat of war. To that end, the symposium concluded by issuing a formal statement that appeared three weeks later as a full-page open letter to President Carter and Chairman Brezhnev in the *New York Times*. It reviewed the dangers the conference had defined, asked that the threat of the superpowers' nuclear capability be recognized, and urged that the world's leaders begin to dismantle their arsenals. Later that same year, PSR organized several other symposiums. One in New York, sponsored by Columbia's College of Physicians and Surgeons and Yeshiva's Albert Einstein College of Medicine, followed the Harvard–Tufts pattern and used many of the same speakers, with the addition of such people as chief SALT II negotiator Paul Warnke and former Secretary of State Cyrus Vance.[23]

Physicians for Social Responsibility relied on new recruits to spread its message. One of the key figures in the revival of the organization was Helen Caldicott, an Australian-born pediatrician who had joined the anti-nuclear movement after reading the novel *On the Beach* as a Melbourne teenager. In 1979, she resigned her practice and teaching appointment at the Harvard Medical School to become president of the group. Tireless in touring the United States, Caldicott spoke passionately of the need to prevent a cataclysmic nuclear war. Often she showed audiences the film *The Last Epidemic*, which described in detail how San Francisco would fare in an atomic attack. If you love this planet, she told listeners, change the priorities in your life. For her, this work simply extended the practice of medicine. "It is the ultimate form of preventive medicine," she said. "If you have a disease and there is no cure for it, you work on prevention."[24]

PSR attracted a tremendous following. As a result of Caldicott's efforts, the organization attracted more than three hundred new recruits a week and membership grew from three thousand in 1981 to sixteen thousand in 1982. Its staff of twenty-four full-time employees worked with a budget that increased from half a million to more than a million and a half dollars in the course of a year. It promoted day-long conferences in cities around the country to spread the message as widely as possible and even spawned an international arm.

Formation of the international group reflected an awareness that the nuclear problem was worldwide. Near the end of 1980, three American and three Russian physicians met in Geneva to consider how to deal with the issue of atomic war in a global context. Two years later, the resulting organization, International Physicians for the Prevention of Nuclear War, included thirty thousand members in thirty countries. Its report, *Last Aid: The Medical Dimensions of Nuclear War*, with contributions by British, Soviet, Japanese, and American physicians and scholars, extended the discussion with attention to the psychological side as well as the physical impact of an attack. H. Jack Geiger, a specialist in community medicine, noted that "for individuals biological survival is possible; for human populations, survival is a social as well as a biological phenomenon, and the ultimate wound is the rupture of the social fabric," for which medical care was nonexistent. Psychiatrist Robert Jay Lifton and sociologist Kai Erikson recalled Nikita Khrushchev's prediction that survivors of an atomic attack would envy the dead and countered that "they would be incapable of such feelings. They would not so much envy as, inwardly and outwardly, resemble the dead," for they could not avoid what Lifton called "psychic numbing." The international physicians' group also provided a televised account of the dangers of nuclear war that was shown to Soviet viewers in June 1982 and to Americans in October of that same year.[25]

PSR and its offshoots revived the approach scientists had taken in the first years of the atomic age. It was an effort to shape opinion and so direct the nation and the world toward control of nuclear weapons. As George Kistiakowsky wrote in an essay in the *Bulletin of the Atomic Scientists* just before his death at the end of 1982, "Forget the channels. There is simply not enough time left before the world explodes. Concentrate instead on organizing, with so many others who are of like mind, a mass movement for peace such as there has not been before."[26]

Other groups besides PSR were equally concerned with the hazards of nuclear war. The Union of Concerned Scientists was one of the most active. Henry W. Kendall, a physicist at the Massachusetts Institute of Technology and former consultant to the Defense Department, had founded the organization in 1969 to oppose the drift toward anti-ballistic missile systems. After winning that battle with the ratification of the SALT I Treaty, the group had shifted focus. The average citizen, Kendall explained, was discouraged by the "forbidding complexity of nuclear arms," which had "been hidden behind the combined shroud of technology and national security." As more and more reactors went on-line and safety concerns mounted, the Union of Concerned Scientists turned to the more popular question of how to control nuclear power. Its membership had reached a hundred thousand by the time of the accident at Three Mile Island in 1979, yet Kendall felt that power was simply a tangential issue and pressed in 1980 to return the organization to its original weapons-oriented approach. He was troubled by the hard-line stance he perceived at the top levels of government and noted how the President's rhetoric frightened others as well. "There was Reagan talking about fighting and winning a limited nuclear war and handing out his laundry list of building up every conceivable nuclear weapon because he claimed we were behind the Russians," Kendall said. "It brought out the latent anxiety." Capitalizing on that discontent, the Union of Concerned Scientists planned a series of teach-ins at colleges on Veterans Day in 1981. Two dozen such assemblies were expected; a hundred fifty were held. The media proved interested, and the attention received from newspapers and television helped the movement along.[27]

At about the same time, direct-action demonstrators in California shifted their focus from nuclear power to nuclear weapons. While several of those having taken part in the protests at the Diablo Canyon plant (see Chapter 6) were still in jail, activists decided to take on the Lawrence Livermore Laboratory at the University of California since it served as a nuclear weapons research facility. The Livermore Action Group, operating out of Berkeley, launched a series of blockades at the laboratory and quickly became the most aggressive disarmament organization in the Bay Area. Its protests over the next several years helped to raise public awareness of the perils of the arms race.[28]

A quieter, but equally incisive, protest came from Catholic bishops in the United States. In May 1983, after extensive discussion and debate, they issued "The Challenge of Peace: God's Promise and Our Response"—a pastoral letter on war and peace. In it, they called "the arms race one of the greatest curses on the human race" and declared that "under no circumstances may nuclear weapons or other instruments of mass slaughter be used for the purpose of destroying population centers or other predominantly civilian targets." It was an articulate effort to mobilize the powers of the church for social ends.[29]

Meanwhile, a number of scientists drew worldwide attention with their hypothesis of a "nuclear winter" that might follow an atomic war. In 1983, nationally known astronomer Carl Sagan and four of his colleagues speculated on the basis of sophisticated computer models that the dust and smoke in the atmosphere following the widespread use of nuclear weapons could produce previously unanticipated results. Sagan had been interested in data transmitted from Mars in late 1971 by the *Mariner* 9 spacecraft. With a global dust storm in progress, scientists could measure the cooling when sunlight was unable to reach the planet's surface and the time it took for normal conditions to return when the storm dissipated. Later he and James B. Pollack and Brian Toon of the National Aeronautical and Space Administration's Ames Research

Center used models developed to investigate volcanic eruptions to apply the Mars findings to Earth. Collaboration with Richard Turco, a scholar at a research and development institute in Marina del Rey, California, who had long been concerned with the effects of nuclear weapons, turned their attention toward the climatic impact of nuclear war.[30]

The group presented its findings at a conference in Washington, D.C., published them in the prestigious journal *Science*, and made them even more accessible with an article Sagan authored in Sunday-supplement *Parade* magazine. "Even small nuclear wars can have devastating climatic effects," Sagan wrote, "enough to generate an epoch of cold and dark." The chaos would be even worse than expected. Numerous plant and animal species would disappear. Vast numbers of surviving humans would starve to death, "and there seems to be a real possibility of the extinction of the human species." Like other activist scientists, Sagan told readers that "fortunately, it is not yet too late. We can safeguard the planetary civilization and the human family if we so choose. There is no more important or urgent issue." Then he pleaded for an international agreement to begin destroying warheads, another compact to reduce the number of nuclear "triggers," and acceptance of a nuclear freeze.[31]

The nuclear freeze was one response to the problems scientists and other critics were defining. It was the idea of Randall Forsberg, first a staffer at the Stockholm International Peace Research Institute and then head of her own Institute for Defense and Disarmament Studies in Brookline, Massachusetts. She initially became interested in the nuclear arms race when she worked as a typist at the Swedish institute and began to read what she typed. Appalled that the 1963 negotiations for a comprehensive test-ban treaty had broken down when the United States insisted on seven yearly inspections and the Soviet Union would allow no more than three, she wondered innocently why compromise was not possible at five. With no answers forthcoming, she began writing her own assessments of the arms race. She came to the notion of a freeze by reflecting further about the inability of the superpowers to achieve a comprehensive settlement. Following that failure, the experts simply accepted the idea of a permanent arms race and dedicated themselves to keeping things equal. "The buzz word," she said, "was stability." But then as Reagan launched a massive military buildup, complete with new missile systems and submarines, the United States seemed to reject even that limited goal.[32]

Forsberg's solution was a mutual and verifiable freeze. It was a simple enough concept for the public to accept easily and, if adopted, could limit the ever-increasing supply of nuclear arms. In early 1980, the Fellowship of Reconciliation organized a meeting of several dozen peace groups to consider the idea. Word reached Vermont, and as a result of the efforts of the American Friends Service Committee, the Forsberg proposal received a favorable hearing in town meetings throughout the state. Such discussions spread elsewhere and were soon occurring around the country. In 1982, several hundred thousand backers marched to Central Park in New York City to send a simple message to national leaders: Do something! Physicians for Social Responsibility embraced the freeze. Ground Zero Week in the spring of that year provided further support as thousands of people in a hundred fifty cities and five hundred communities dramatized the devastating effects of a nuclear war. At a national political level, Senator Edward M. Kennedy (a Democrat from Massachusetts) and Senator Mark O. Hatfield (a Republican from Oregon) introduced a joint congressional resolution calling for a weapons freeze. They countered the Reagan administration's talk of a "window of vulnerability" with the assertion that there was rather a "window of opportunity" for arms control. They proposed, in short, to create a "firebreak" to circumscribe the arms race.[33]

Not everyone agreed that a freeze was the answer. Christopher M. Lehman, director of the Office of Strategic Nuclear Policy in the State Department, argued that fear—according to him, the guiding element in the freeze campaign—was not enough to prevent nuclear war. He claimed that while the freeze was superficially attractive, it would undermine the policy of deterrence, which had helped the world avoid an atomic catastrophe in the post–World War II years. When Ariela Gross, a seventeen-year-old high school student from Princeton, New Jersey, was selected as a Presidential Scholar and asked other recipients of the prestigious prize to join her in signing a petition to the President supporting the freeze, the administration reacted with anger and threatened to withdraw her award. Responding to the messages coming from high-ranking officials, humorist Russell Baker declared, "My position on the nuclear freeze is that the government ought to stop telling me I'm too dumb to have an opinion on it."[34]

After introduction of the Kennedy–Hatfield resolution in Congress, two dozen senators and more than a hundred fifty representatives signed their names as sponsors. In August 1982, the House of Representatives defeated the measure by a 204 to 202 vote but later passed it. In the November elections, nine states and many more communities supported the proposal. While it never became national policy, it clearly created pressure on the administration to consider more seriously the effort to control arms.[35]

Vigorously opposed to the freeze, Reagan responded by proposing a massive program to bolster American defense. After finally acknowledging the pervasive fear of nuclear war midway through his first term, he searched for some way to counter that anxiety while maintaining his administration's massive arms buildup. His Strategic Defense Initiative or Star Wars plan, with its call for a space-based nuclear umbrella, offered a quick fix to political problems by promising absolute security. In one fell swoop, the President undercut the freeze by appropriating its rhetoric and shifted the terms of the debate.[36]

Like other Americans, Reagan had long fantasized about special weapons that could make the nation safe from any attack. Years before, just after the outbreak of World War II, he had played Secret Service agent Brass Bancroft in the film *Murder in the Air*. His mission was to defend a new superweapon, a "death-ray projector" that could stop enemy planes, make America invincible, and so serve as "the greatest force for world peace ever discovered." That background may well have affected the thinking behind the Star Wars plan.[37]

A more immediate impetus came from Edward Teller, the brilliant physicist who had played a major part in developing the hydrogen bomb. For decades, Teller had dreamed of a new generation of strategic weapons that could focus the energy from small nuclear explosions and provide a more effective means of missile defense. In 1967, he shared his ideas with Reagan, the newly elected governor of California, who was on a tour of the Lawrence Livermore Laboratory southeast of San Francisco. Fifteen years later, with Reagan in the White House, Teller made another pitch. Accompanied by three of Reagan's close friends—businessman

Karl Bendetsen, brewer Joseph Coors, and rancher/oilman William Wilson—he met the President after a Cabinet session. Together the group urged Reagan to authorize a massive program like the Manhattan Project to develop new weapons to counter the Soviet threat. In subsequent meetings during the next year, Teller promoted a secret research project at Livermore known as "Excalibur"—a nuclear-bomb-pumped X-ray laser which could shoot down Russian ICBMs—and warned that the Soviets might get such a weapon first unless the United States embarked upon a serious developmental effort. He also lobbied the Joint Chiefs of Staff, members of Congress, and senior Defense Department officials to support such a program.[38]

Reagan was receptive. On March 23, 1983, he outlined his vision for a new kind of nuclear shield in the course of a televised speech on Moscow's arms buildup and interest in Central America. A number of Manhattan Project veterans had been invited to the White House to listen to the speech in person, and, just before he began, the President told Teller, "Edward, you're going to like it." At the end of his speech, Reagan shifted gears and posed a rhetorical question: "Wouldn't it be better to save lives than to avenge them?" Then he unveiled his own ideas for a novel strategic approach. "What if free people," he asked, "could live secure in the knowledge that their security did not rest upon the threat of instant U.S. retaliation to deter a Soviet attack, that we could intercept and destroy strategic ballistic missiles before they reached our own soil or that of our allies?" The means to attain such security involved an enormous, long-term defensive program to develop a space-based shield that would be so effective it would render nuclear weapons obsolete and thereby "free the world from the threat of nuclear war." Reagan's approach, which was admittedly political and took many of his own technical advisers by surprise, provided an altogether new focus for discussion of the nuclear threat.[39]

Over the next few years, scientists and administration officials elaborated on the details of Reagan's vision. Air Force lieutenant general James Abrahamson, once an astronaut hopeful and later director of the space-shuttle program, headed the effort from the Pentagon. The chief scientist was Gerald Yonas, who turned to weapons research after working on the question of fusion power, and he was assisted by a host of other scientists at the nation's three major weapons laboratories—Lawrence Livermore in California and Los Alamos and Sandia in New Mexico—who welcomed the prospect of spending thirty billion dollars on the project in the next five years. Some speculated about the possibility of creating chemical lasers, with chemical reactions emitting infrared radiation that could then be amplified and aimed at enemy targets. Teller continued to promote Excalibur, his X-ray laser, which could be deployed in similar ways. Other scientists proposed using particle beams, which were made up of streams of atomic or subatomic particles rather than light waves, to knock out incoming targets. Still others proposed a series of orbiting mirrors to focus the beams generated in different ways. Yonas himself considered what came to be called the Jedi Concept—a plan for firing plasma globs, made up of energized nuclei and electrons, into space at close to the speed of light, if only a way could be found to make the plasma stick together. The proposed Strategic Defense Initiative was an enormous undertaking, dwarfing even the Manhattan Project, but, as opponents pointed out, was based on technology that was far less mature and sometimes did not even exist.[40]

As the research effort gained momentum, critics lashed out at the entire scheme. Was the goal of rendering nuclear weapons obsolete a desirable one, they asked, or did such weapons help prevent war between the superpowers? Was the aim realistic and attainable? Would the preliminary research effort generate such momentum among scientists, contractors, and members of Congress that it might never be stopped?[41]

The scientific community was split, as it had been in the debate over the hydrogen bomb, and soon found itself embroiled in an even more bitter debate. Scientists opposed to the Star Wars program were among those most active in attacking the position of their more sympathetic colleagues and framing the public critique. A year and a half after Reagan's speech, the Union of Concerned Scientists published a book entitled *The Fallacy of Star Wars*, which marshaled arguments for use by opposition forces. Carl Sagan, writing again in *Parade* magazine, used a birth control analogy to argue that the program would not work, for it could not be absolutely effective: "A contraceptive shield that deters 90 percent of 200 million sperm cells is generally considered worthless—20 million sperm cells penetrating the shield are more than enough. Such a shield is *not* better than nothing; it is worse than nothing, because it might well engender a false sense of security, bringing on the very event it was designed to prevent. The same is true for the leaky shield of Star Wars." Richard Garwin, a physicist who had helped develop the hydrogen bomb, had later served as Richard Nixon's science adviser, and who now worked for IBM, suggested that computer technology was not yet prepared to monitor such a system. "The computer would require 10 million lines of error-free code," he said. "I don't know anyone who knows that that is possible." Kosta Tsipis, a research scientist from the Massachusetts Institute of Technology, spoke for many of his associates in calling the program "a technological science-fiction fantasy." About a thousand scientists around the country signed a pledge not to participate in the Star Wars effort. Sheldon L. Glashow, a Nobel Prize–winning physicist from Harvard, was among those voicing his opposition, even though the theoretical research he did hardly fed into the program. "I would give 'Star Wars' a 'D,' " he said, "because it is a danger to peace, a dis-inclination to arms control, deleterious to American science, and it is destabilizing, dumb, and damned expensive."[42]

But scientists were not alone in attacking the plan. Senator Edward Kennedy called it a "grand illusion." Gerard Smith, chief negotiator of the SALT I Treaty, claimed that the earlier agreement banned such anti-ballistic missile projects and deplored the administration's effort to reinterpret the ABM accord to make the Star Wars program possible. Political psychologist Lloyd Etheridge spoke of the mythic appeal of the effort, consonant with Reagan's general approach, which overrode mundane considerations about whether or not it would work. It was, according to Harvard physician John E. Mack, academic director of the Center for Psychological Studies in the Nuclear Age, "nationalism as theater."[43]

The debate was played out in the media. When Ronald Reagan proved displeased by the term "Star Wars"—used by Edward Kennedy on the floor of the Senate the day after the 1983 speech but apparently employed even earlier by a number of journalists speculating about space-weapons technology—commentator William Safire published his readers' suggestions for alternative names in the *New York Times Magazine*. Out of six hundred responses to an appeal, he reported such acronyms as: BONZO—Ballistic Offense Neutralization Zone or Bulwark Order Negating Zealous Offensive; DUMB—Defensive Umbrella; WACKO—Wistful Attempts to Circumvent Killing Ourselves; and WIMP—Western Intercontinental Missile Protection. Proponents, to be sure, came up with more sympathetic terms that he also recorded: DEUS—Defense of Upper Space; SAFE—Shield against Fatal Encounter; DISARM—Defense in Space against Russian Missiles; and HOPE—Hostile Projectile Elimination. The issue was joined as well in the world of television commercials. The Union of Concerned Scientists produced an ad showing a little boy in his pajamas, holding a teddy bear, looking out into the sky from his bedroom window. As he sang "Twinkle, Twinkle, Little Star," one star brightened, then blew up in his face. At that point the voice of Darth Vader, the black presence in the *Star Wars* trilogy, declared, "Stop Star Wars, stop weapons in space." In response, a Coalition for the Strategic Defense Initiative released its own commercial which showed a child's crayon drawing of a family standing outside its house and used the voice of a young girl to say: "I asked my daddy what this 'Star Wars' stuff is all about. He said right now we can't protect ourselves from nuclear weapons, and that's why the president wants to build a peace shield. It'd stop missiles in outer space . . . so they couldn't hit our house. Then nobody could win a war, and if nobody could win a war, there's no reason to start one. My daddy's smart."[44]

The debate clarified the issues but failed to provide a resolution. Even as proponents plunged ahead with their research plans, confusion remained about what precisely might be possible. As columnist Art Buchwald observed, "The beauty of the Star Wars defense system is that everyone can discuss it with authority, because no one, including the people in charge, has any idea of what it is." Worse than the confusion was the anxiety that persisted. The clock on the cover of the *Bulletin of the Atomic Scientists*—that barometer of the arms race—moved ahead to three minutes before midnight in recognition of the unwillingness of the superpowers to negotiate about ever more pressing concerns.[45]

Near the end of the Reagan presidency, interest in the Strategic Defense Initiative faded. Polls revealed that a majority of Americans were ambivalent about Reagan's vision, and political support began to decline. As reports leaked out about turmoil in the weapons labs and rumors surfaced that Edward Teller had misled the President—and the public—about the possibility of producing an X-ray laser, Star Wars lost its luster. The resumption of arms-control talks undermined the effort and led a new chief executive, George Bush, to seek less funding and a new Congress to view all requests less sympathetically than before. By the 1990s, the program was but a shadow of Reagan's dream.[46]

The protest against Star Wars, and against burgeoning nuclear arsenals, made a difference. Activists—scientists and nonscientists alike—were able to dramatize their concerns and to mobilize public opinion more effectively than ever before. In the process, they forced institutional leaders to listen and pressured the reluctant administration to resume the process of negotiation it had chosen to ignore. Scientists were involved in this effort, as they had been in past efforts, but they hardly worked alone. Indeed, their greatest contribution was in stirring others throughout the society to join them in what became the most forceful protest movement in the atomic age. Whether focusing on the likely consequences of a nuclear war or challenging the misguided efforts to create a foolproof means of defense, these protest groups, working together, helped create a national consensus that something had to be done. Results did not occur overnight, to be sure. But, as Bernard Lown noted after watching public concern develop over several decades, "It's like boiling water. Nothing happens, nothing happens, nothing happens, and then finally there's steam." Eventually, as Robert Jay Lifton observed in quoting poet Theodore Roethke at the first PSR symposium, "In a dark time, the eye begins to see."[47]

Epilogue

Protest against unbridled nuclear development has occasionally forced a shift in America's foreign and domestic policy in the first fifty years of the atomic age. Once a decade, in diplomatic affairs, the United States has gingerly backed away from its most truculent positions and sought accommodation with adversaries in a joint effort to mute the global threat of nuclear conflagration. But each time that venture has gone only so far in confronting pressing problems of the proliferation and potentially deadly use of nuclear weapons, and the initiative has faded from full view without ending the worst perils. Meanwhile, on the home front, technology has improved and led to even more extravagant versions of the arms race. Why, in the face of escalating stakes, have we failed to go further in addressing the real possibility of atomic doom? And why have we been unable to bring about long-sought lasting change in the nation's commitment to nuclear development?

Answers to those questions lie with the very nature of the public dialogue about atomic issues in the years since 1945. The triangular conversation that has included scientists, commentators, and policymakers has allowed various voices to speak loudly and expressively. The scientists and their more creative counterparts in the world of fiction and film who have warned audiences about the dangers of nuclear holocaust might have managed from time to time to capture the headlines and dominate the news. But the dialogue has provided no natural mechanism to force a response or shift the government's course. Only when critics have been able to focus on prickly political problems and cause enough of a public scare have they been able to attract the attention of top officials. And even then they have taken but the first steps toward encouraging those government leaders to alter their stance.

Today, for the first time since the end of World War II, the scenario has changed. The conversation continues with the same parties participating, but the framework is different. The terms of the debate—and the very dialogue itself—have shifted dramatically in response to world events. With the precipitous end to the Cold War, the prospect of a deadly nuclear war with the Soviet Union has diminished. And yet, while the world may have a better chance today than at any time over the past century to secure stability in the atomic realm, there are still powerful problems that cannot be ignored. We have to worry about the fate of nuclear weapons in the former Soviet republics. We need to confront the continuing question of proliferation, particularly in the Middle East. And, with a heightened sense of the destruction we have wrought on our planet, we need to deal with issues of nuclear waste that will affect our environment for centuries to come.

The real question as we look ahead is how to break out of the pattern that has dominated the nuclear dialogue in the past half century. After the dramatic demonstration of atomic might at Hiroshima and Nagasaki, most Americans responded to the nuclear age with a cheery optimism about the brave new world to come. Only in time did lurking fears become more powerful and crystallize in the criticism of the nuclear plans that were unfolding. We are, of course, more cognizant today of the pitfalls ahead than we were in 1945, but we must still guard against the wishful thinking that the end of the Cold War alone is enough to make the planet a safer place. We need to address continuing problems, even as we savor the possibilities of a more peaceful world.

As we move forward, it is instructive to remember the main lesson of the past fifty years: that public awareness of problems is the key to potential change. At repeated intervals in the post–World War II years, the American public has reacted, often aggressively, when it has grasped potential dangers, and that reaction has catalyzed significant policy shifts. But just how does the public learn what is happening today?

Keeping the public informed is the dilemma of modern democracy. A century ago, although women and members of a number of minority groups were denied access to the public sphere, many members of the, albeit limited, electorate read newspapers, debated issues publicly, and generally were active participants in the political process. With the necessary expansion of the electorate, we have had to face the continuing problem of how to generate mass involvement in the world of national affairs. That process has become harder still with changes in the way we learn what is going on. In the first decade following World War II, members of the news media and the literary elite were instrumental in creating informed public opinion and alerting Americans to the dangers that lay ahead. But in the past several decades, television has eroded interest in reading and changed the way the public learns about current affairs. Television has become an increasingly important part of the political process, and ubiquitous sound bites have taken the place of the written accounts that once helped frame questions for debate. Perhaps as a result of a less reasoned effort to deal with the intricacies of issues, interest in the political process has declined.

In the postwar years, public relations have played an ever larger role in shaping national policy. Following World War I, many Americans worried about the power of propaganda, especially after George Creel's Committee on Public Information stirred up hatred of all things German, and their anxieties circumscribed the formal propaganda effort in World War II. Nonetheless, skillful national leaders learned the most effective ways to mobilize support for political ends. Franklin Roosevelt was a master at using the radio to generate public confidence. Dwight Eisenhower played upon his avuncular image in his dealings with the press to orchestrate national support. And Ronald Reagan was equally adept in front of television cameras as he sought to communicate his vision of where the nation should go.

Responding to those successful initiatives, some nuclear critics experimented with the same techniques. They were impressed with the way film director Stanley Kubrick conveyed his sense of absurdity in *Dr. Strangelove or: How I Learned to Stop Worrying and Love the Bomb*. Some, like novelist Tim O'Brien in *The Nuclear Age*, employed fiction to convey the haunting fears of the 1950s. Others, like physician Helen Caldicott, used television and film to spread their message around the world. Still other activists, like supporters of the nuclear freeze, organized demonstrations that caught the attention of television journalists and were shared with the public around the nation and, with the advent of the Cable News Network (CNN), around the globe. The print world was not entirely abandoned. Jonathon Schell demonstrated the continuing power of the written word in *The Fate of the Earth*. But activists were now willing to rely on new, and potentially more powerful, techniques.

These efforts have met with some success but need to be coupled with further grass-roots activity to mobilize and educate the public. It *is* necessary for Americans to remain informed, and to press for disclosure about nuclear intentions, even when leaders from both political parties resist sharing their plans. The bomb may have become, as physicist Alvin C. Graves observed in 1952, a fact of life, like a heart condition, that has to be lived with. The arms race may have become a kind of ritual, a means of taking one step after another according to a long-settled script. But there *are* ways of changing the dialogue, as the upheaval in eastern Europe demonstrates, if only the people can remain alert to developments around them and willing to press on until change occurs.[1]

The conversation may occasionally turn shrill, but that is the price of democracy. The nuclear genie is out of the bottle—as Walt Disney observed decades ago—and can never be returned to its former resting place. But, as activists have argued, we *can* change our fate and avoid the catastrophe we have made possible with our own hands. Environmentalists, fearful of the lethal effects of nuclear waste, have shown how communities can be mobilized and lawsuits can be won to force necessary change. Bernard Lown, so effective in organizing Physicians for Social Responsibility, expressed a commonly held position best of all. "We are but transient passengers on this planet Earth," he told audiences in the 1980s. "It does not belong to us. We are not free to doom generations yet unborn. We are not at liberty to erase humanity's past or dim its future." Like other doctors, he was "aware of the resiliency, courage, and creativeness that human beings possess" and shared "an abiding faith in the concept that humanity can control what humanity creates." His skill in shaping opinion, spreading a message, and making governments around the world respond demonstrates the possibilities of dealing with the most ticklish issues of the nuclear age. For while activists have not always had their way, at key points in the past they have been able to break the bureaucratic logjam and promote a more reasonable approach. Their success, however modest, offers the last best hope for the future.[2]

Notes

Prologue

1. Nicholas Humphrey, "The Bronowski Memorial Lecture: 'Four Minutes to Midnight,' " *The Listener*, 29 Oct. 1981, 493. The lecture is reprinted in Nicholas Humphrey, *Consciousness Regained: Chapters in the Development of Mind* (New York: Oxford Univ. Press, 1984), 194–209.

2. Henry L. Stimson, "The Decision to Use the Atomic Bomb," *Harper's Magazine*, Feb. 1947, 107; Lewis Thomas quoted in John E. Mack, " 'But What About the Russians?': A Psychiatrist Looks at American Nuclear Arms Paranoia," *Harvard Magazine*, Mar.–Apr. 1982, 21.

3. Robert M. Hutchins quoted in Paul Boyer, *By the Bomb's Early Light: American Thought and Culture at the Dawn of the Atomic Age* (New York: Pantheon, 1985), 112; Heinz Haber, *The Walt Disney Story of Our Friend the Atom* (New York: Simon & Schuster, 1956), 13.

4. Hazel Gaudet Erskine, "The Polls: Atomic Weapons and Nuclear Energy," *Public Opinion Quarterly* 27 (Summer 1963): 157, 160.

5. I. I. Rabi quoted in Roger Rosenblatt, "The Atomic Age," *Time*, 29 July 1985, 33; Isaac Asimov, *Opus 100* (Boston: Houghton Mifflin, 1969), 105.

1. Origins of the Atomic Age

1. Frederick Soddy quoted in Spencer R. Weart, *Nuclear Fear: A History of Images* (Cambridge, Mass.: Harvard Univ. Press, 1988), 6.

2. H. G. Wells, *The World Set Free* (New York: E.P. Dutton, 1914), 114–17, 152, 221–22.

3. Weart, *Nuclear Fear*, 78–79; Carol S. Gruber, "Manhattan Project Maverick: The Case of Leo Szilard," *Prologue* 15 (Summer 1983): 73–75, 83; Szilard quoted in Martin J. Sherwin, *A World Destroyed: The Atomic Bomb and the Grand Alliance* (New York: Alfred A. Knopf, 1975), 21; Rutherford quotation and other background in Richard Rhodes, *The Making of the Atomic Bomb* (New York: Simon & Schuster, 1986), 13–28.

4. Richard G. Hewlett and Oscar E. Anderson, Jr., *The New World, 1939–1946*, Vol. 1 of *A History of the United States Atomic Energy Commission* (Wash-

ington, D.C.: U.S. Atomic Energy Commission, 1972), 10–11; Sherwin, *A World Destroyed*, 14–17; Weart, *Nuclear Fear*, 77.

5. Laura Fermi quoted in Rhodes, *The Making of the Atomic Bomb*, 207; Daniel J. Kevles, *The Physicists: The History of a Scientific Community in Modern America* (New York: Vintage, 1979), 324; Sherwin, *A World Destroyed*, 17.

6. *Berliner Illustrirte Zeitung* quoted in Rhodes, *The Making of the Atomic Bomb*, 168.

7. Albert Einstein to F. D. Roosevelt, 2 Aug. 1939, President's Secretary's File: Safe: Alexander Sachs, Franklin D. Roosevelt Library, Hyde Park, New York; for a facsimile of this letter, and numerous other documents pertaining to the development of the bomb, see Michael B. Stoff, Jonathan F. Fanton, and R. Hal Williams, eds., *The Manhattan Project: A Documentary Introduction to the Atomic Age* (New York: McGraw-Hill, 1991); This letter and other documents also appear in Robert C. Williams and Philip L. Cantelon, eds., *The American Atom: A Documentary History of Nuclear Policies from the Discovery of Fission to the Present, 1939–1984* (Philadelphia: Univ. of Pennsylvania Press, 1984).

8. Roosevelt quoted in Hewlett and Anderson, *The New World*, 17; Sherwin, *A World Destroyed*, 27–31, 36.

9. Sherwin, *A World Destroyed*, 38–39, 42.

10. Ibid., 42; See also F. G. Gosling, *The Manhattan Project: Science in the Second World War* (Washington, D.C.: History Division, Executive Secretariat, Office of Administration and Human Resources Management, U.S. Department of Energy, 1990), 19–36.

11. Kenneth D. Nichols quoted in Rhodes, *The Making of the Atomic Bomb*, 426; Robert Jungk, *Brighter Than a Thousand Suns: A Personal History of the Atomic Scientists* (New York: Harcourt Brace, 1958), 117.

12. Hewlett and Anderson, *The New World*, 76–83; Vincent C. Jones, *Manhattan: The Army and the Atomic Bomb, Special Studies: United States Army in World War II* (Washington, D.C.: Center of Military History, U.S. Army, 1985), 57–61, 80–82; Sherwin, *A World Destroyed*, 53–63.

13. James Phinney Baxter, 3rd, *Scientists against Time* (Cambridge, Mass.: MIT Press, 1968), 3; Sherwin, *A World Destroyed*, 51.

14. Albert Wattenberg, "December 2, 1942: The Event and the People," *Bulletin of the Atomic Scientists* 38 (Dec. 1982): 22–32; Len Ackland, "Dawn of the Atomic Age," *Chicago Tribune Magazine*, 28 Nov. 1982, 10, 12, 19–23.

15. Hewlett and Anderson, *The New World*, 84–103.

16. Nuel Pharr Davis, *Lawrence & Oppenheimer* (New York: Simon & Schuster, 1968), 20–25; Emilio Segrè and Leslie Groves quoted in Rhodes *The Making of the Atomic Bomb*, 444, 448; United States Atomic Energy Commission, *In the Matter of J. Robert Oppenheimer: Transcript of Hearing before Personnel Security Board, Washington, D.C., April 12, 1954 through May 6, 1954* (Washington, D.C.: U.S. Government Printing Office, 1954), 7–11.

17. Sherwin, *A World Destroyed*, 53–54; Emilio Segrè and Leo Szilard quoted in Rhodes, *The Making of the Atomic Bomb*, 451–53; Alice Kimball Smith and Charles Weiner, eds., *Robert Oppenheimer: Letters and Recollections* (Cambridge, Mass.: Harvard Univ. Press, 1980), 242–46.

18. I. I. Rabi quoted in Rosenblatt, "The Atomic Age," 33; David Gates,

"Thinking About the Apocalypse," *Newsweek*, 2 Mar. 1987, 72; Paul Olum, "Why Did We Work to Build Such a Terrible Thing?," *The Oregonian*, 4 Aug. 1985.

19. John J. Weltman, "Trinity: The Weapons Scientists and the Nuclear Age," *SAIS Review* 5 (Summer–Fall 1985): 30–31; Rhodes, *The Making of the Atomic Bomb*, 543–45.

20. Weltman, "Trinity," 29.

21. Rhodes, *The Making of the Atomic Bomb*, 656, 668, 676; quotations in James Gleick, "After the Bomb, a Mushroom Cloud of Metaphors," *New York Times Book Review*, 21 May 1989, 53.

22. Harry Truman quoted in Robert J. Donovan, *Conflict and Crisis: The Presidency of Harry S Truman, 1945–1948* (New York: W.W. Norton, 1977), 15, 17; Allan M. Winkler, "Harry S Truman and Leadership in the Atomic Age," in David M. Kennedy and Michael E. Parrish, eds., *Power and Responsibility: Case Studies in American Leadership* (San Diego: Harcourt Brace Jovanovich, 1986), 67–69.

23. Harry S Truman, *Memoirs*, Vol. 1, *Year of Decisions* (Garden City, N.Y.: Doubleday, 1955), 10.

24. Ibid., 10; Henry L. Stimson, memorandum discussed with the President, 25 Apr. 1945, in Diary of Henry L. Stimson, Henry L. Stimson Papers, Manuscripts and Archives, Yale Univ. Library, New Haven, Conn.

25. Barton J. Bernstein, "Roosevelt, Truman, and the Atomic Bomb, 1941–1945: A Reinterpretation," *Political Science Quarterly* 90 (Spring 1975): 36–37; Michael S. Sherry, *The Rise of American Air Power: The Creation of Armageddon* (New Haven, Conn.: Yale Univ. Press, 1987), 316–17; Harry Truman quoted in Donovan, *Conflict and Crisis*, 67.

26. Leo Szilard quoted in Rhodes, *The Making of the Atomic Bomb*, 636–37.

27. Sherwin, *A World Destroyed*, 210–13.

28. Leo Szilard quoted in Paul Bracken, "Books," *International Herald-Tribune*, 19 Dec. 1984; Scientific Panel, Recommendations on the Immediate Use of Nuclear Weapons, 16 June, 1945, in Sherwin, *A World Destroyed*, Appendix M, 304–5; Peter Wyden, *Day One: Before Hiroshima and After* (New York: Simon & Schuster, 1984) 176–80.

29. Harry Truman quoted in Robert L. Messer, "New Evidence on Truman's Decision," *Bulletin of the Atomic Scientists* 41 (Aug. 1985): 54–55; Paul Boyer, " 'Some Sort of Peace': President Truman, the American People, and the Atomic Bomb," in Michael J. Lacey, ed., *The Truman Presidency* (New York: Woodrow Wilson International Center for Scholars and Cambridge Univ. Press, 1989), 181; Donovan, *Conflict and Crisis*, 93.

30. Bernstein, "Roosevelt, Truman, and the Atomic Bomb," 51; Sherwin, *A World Destroyed*, 228–29; Ernie Pyle, *Ernie's War: The Best of Ernie Pyle's World War II Dispatches*, edited by David Nichols (New York: Simon & Schuster, 1986), 367.

31. Diary of Henry L. Stimson, 15 May 1945, Henry L. Stimson Papers; Sherwin *A World Destroyed*, 197–98; Hewlett and Anderson, *The New World*, 391.

32. Sherwin, *A World Destroyed*, 135–36, 203; Gabriel Kolko, *The Politics*

of War: The World and United States Foreign Policy, 1943–1945 (New York: Random House, 1968), 541–43; Donovan, *Conflict and Crisis*, 97; Leslie R. Groves, *Now It Can Be Told: The Story of the Manhattan Project* (New York: Harper & Row, 1962), 265.

33. George F. Will, "An Exhibit: *Enola Gay* for the Kids," *International Herald-Tribune*, 20–21 July 1985.

34. Masuku Ibuse, *Black Rain*, translated by John Bester (Palo Alto, Calif.: Kodansha International Ltd., 1969), 12.

35. Ibid., 37, 54–55.

36. *Unforgettable Fire: Pictures Drawn by Atomic Bomb Survivors*, edited by the Japanese Broadcasting Corporation (New York: Pantheon, 1977), 7–8; Ibuse, *Black Rain*, 118; "Tatsue Urata's Story," in Takashi Nagai, ed., *We of Nagasaki: The Story of Survivors in an Atomic Wasteland*, translated by Ichiro Shirato and Herbert B. L. Silverman (New York: Duell, Sloan and Pearce, 1951), 82; Donald N. Michael, "Civilian Behavior under Atomic Bombardment," *Bulletin of the Atomic Scientists* 11 (May 1955): 175.

37. *Unforgettable Fire*, 8, 54, 55; Ibuse, *Black Rain*, 121, 188; "Matsu Moriuchi's Story," in Nagai, *We of Nagasaki*, 124; John Hersey, *Hiroshima* (New York: Bantam, 1966), 53, 75. For other, more recent, descriptions of the bombs, see Rosenblatt, "The Atomic Age," 22, 25–28; and "Living with the Bomb: The First Generation of the Atomic Age," *Newsweek*, 29 July 1985, 31, 36–38.

38. Sherwin, *A World Destroyed*, 233–37.

39. Truman, *Memoirs*, Vol. 1, 421–23; Statement by the President of the United States, 6 Aug. 1945, Box 227, President's Secretary File, Papers of Harry S Truman, Harry S Truman Library, Independence, Mo.; Donovan, *Conflict and Crisis*, 96.

40. H. Bruce Franklin, "Fatal Fiction: A Weapon to End All Wars," *Bulletin of the Atomic Scientists* 45 (Nov. 1989): 18, 24; H. Bruce Franklin, *War Stars: The Superweapon and the American Imagination* (New York: Oxford Univ. Press, 1988), 82. Spencer R. Weart, "Learning How to Study Images as Potent Forces in Our History," *Chronicle of Higher Education*, 8 Mar. 1989, A44; A. Merriman Smith, *Thank You, Mr. President: A White House Notebook* (New York: Harper & Brothers, 1946), 286.

41. Harry Truman quoted in Robert H. Ferrell, *Harry S Truman and the Modern American Presidency* (Boston: Little, Brown, 1983), 54; Harold Smith and Harry Truman quoted in Donovan, *Conflict and Crisis*, 97.

42. Robert H. Ferrell, ed., *Off the Record: The Private Papers of Harry S Truman* (New York: Harper & Row, 1980), 304; Harry S Truman, *Memoirs*, Vol. 1, 419; Ferrell, *Harry S Truman*, 56; Margaret Truman, *Harry S Truman* (New York: William Morrow, 1973), 567.

43. Paul Brians, *Nuclear Holocausts: Atomic War in Fiction* (Kent, Ohio: Kent State Univ. Press, 1987), 9, 155–56; H. Bruce Franklin, "Nuclear War and Science Fiction," in H. Bruce Franklin, ed., *Countdown to Midnight* (New York: DAW Books, 1984), 15–16.

44. William L. Laurence, "Nagasaki Was the Climax of the New Mexico Test, *Life*, 24 Sept. 1945, 30; William L. Laurence, "Is Atomic Energy the Key to Our Dreams?," *The Saturday Evening Post*, 13 Apr. 1946, 9–10, 37; Boyer, *By*

the Bomb's Early Light, 117, 187; Paramount newsreel quoted in Weart, *Nuclear Fear*, 104.

45. Paul Fussell, *Thank God for the Atom Bomb and Other Essays* (New York: Summit Books, 1988), 14, 20; William Manchester, *Goodbye, Darkness: A Memoir of the Pacific War* (Boston: Little, Brown, 1980), 210; "Atomic Cafe: Radioactive Rock'n Roll, Blues, Country & Gospel," Rounder Records, n.d.

46. Boyer, *By the Bomb's Early Light*, 10; Joseph C. Goulden, *The Best Years, 1945–1950* (New York: Atheneum, 1976), 261; "Gags Away," *The New Yorker*, 18 Aug. 1945, 16; "Anatomic Bomb," *Life*, 3 Sept. 1945, 53.

47. Richard J. Cushing, "A Spiritual Approach to the Atomic Age," *Bulletin of the Atomic Scientists* 4 (July 1948): 222; Boyer, *By the Bomb's Early Light*, 11; Elaine Tyler May, *Homeward Bound: American Families in the Cold War Era* (New York: Basic Books, 1988), 111; "Atomic Cafe."

48. David E. Lilienthal, *Change, Hope, and the Bomb* (Princeton, N.J.: Princeton Univ. Press, 1963), 18; Goulden, *The Best Years*, 261; Enrico Fermi quoted in Peter Pringle and James Spigelman, *The Nuclear Barons* (New York: Holt, Rinehart and Winston, 1981), 147; *New York Times*, 8 Aug. 1945; Edward Zuckerman, "Atomic Dream & Nightmare," *New York Times Book Review*, 31 Oct. 1982, 9; Laurence, "Is Atomic Energy the Key to Our Dreams?," 10, 41.

49. George H. Gallup, *The Gallup Poll: Public Opinion, 1935–1971*, Vol. 1, *1935–1948* (New York: Random House, 1972), 521–22, 527; Lewis Mumford quoted in Gregg Herken, *The Winning Weapon: The Atomic Bomb in the Cold War, 1945–1950* (New York: Alfred A. Knopf, 1980), 311.

50. H. V. Kaltenborn and Edward R. Murrow quoted in Boyer, *By the Bomb's Early Light*, 5, 7; George Bernard Shaw, " 'What Does Atom Bomb Cost?' Shaw Asks—and Answers," *Washington Post*, 19 Aug. 1945; [Norman Cousins], "Modern Man is Obsolete," *The Saturday Review of Literature*, 18 Aug. 1945, 5; [E. B. White], "Notes and Comment," *The New Yorker*, 18 Aug. 1945, 13.

51. Groves, *Now It Can Be Told*, 296–97; H. A. Bethe, "Can Air or Water Be Exploded?," *Bulletin of the Atomic Scientists* 1 (15 Mar. 1946): 5.

52. Erskine, "*The Polls*," 156; Boyer, *By the Bomb's Early Light*, 22–23.

53. "The War Ends," *Life*, 20 Aug. 1945, 25–31; Peter B. Hales, "The Atomic Sublime," paper presented at the American Studies Association Convention, New York, 23 Nov. 1987, 1–2.

54. Boyer, *By the Bomb's Early Light*, 203–4; Weart, *Nuclear Fear*, 107; Brians, *Nuclear Holocausts*, 4.

55. Hersey, *Hiroshima*, 38, 67; Boyer, *By the Bomb's Early Light*, 205; David Dowling, *Fictions of Nuclear Disaster* (Iowa City, Iowa: Univ. of Iowa Press, 1987), 48.

56. Joseph Luft and W. M. Wheeler, "Reaction to John Hersey's 'Hiroshima,' " *Journal of Social Psychology* 28 (Aug. 1948): 135–40; Charles Poore, " 'The Most Spectacular Explosion in the Time of Man,' " *New York Times Book Review*, 10 Nov. 1946, 56; Ruth Benedict, "The Past and the Future," *The Nation*, 7 Dec. 1946, 656.

57. Sherry, *The Rise of American Air Power*, 153–54, 260, 273–77; Michael Sherry, "The Slide to Total Air War," *The New Republic*, 16 Dec. 1981, 24.

58. Philip Morison and *Chicago Sun* quoted in Boyer, *By the Bomb's Early*

Light, 215; Alexander P. de Seversky, "Atomic Bomb Hysteria," *The Reader's Digest*, Feb. 1946, 121–22.

59. "Editorial," *The Christian Century*, 15 Aug. 1945, 923.

60. Weart, *Nuclear Fear*, 105; Boyer, *By the Bomb's Early Light*, 22.

2. The Question of Control

1. Fraser J. Harbutt, *The Iron Curtain: Churchill, America, and the Origins of the Cold War* (New York: Oxford Univ. Press, 1986), 151–208.

2. Eugene Rabinowitch, "Five Years After," *Bulletin of the Atomic Scientists* 7 (Jan. 1951): 3; Boyer, *By the Bomb's Early Light*, 49.

3. Alice Kimball Smith, *A Peril and a Hope: The Scientists' Movement in America, 1945–1947* (Chicago: The Univ. of Chicago Press, 1965), 528; Alice Kimball Smith, "Scientists and Public Issues," *Bulletin of the Atomic Scientists* 38 (Dec. 1982): 38–39.

4. Albert Einstein quoted in Wyden, *Day One*, 342; Albert Einstein quoted in Howard Morland, "The H-bomb Secret," *The Progressive*, Nov. 1979, 17; Albert Einstein, *Out of My Later Years* (New York: Philosophical Library, 1950), 200, 202; Albert Einstein as told to Raymond Swing, "Einstein on the Atomic Bomb," *Atlantic Monthly*, Nov. 1945, 45; Albert Einstein quoted in Smith and Weiner, *Robert Oppenheimer*, 309; Albert Einstein quoted in David E. Lilienthal, "Democracy and the Atom," *NEA Journal*, Feb. 1948, 80; Albert Einstein, telegram, 23 May 1946, in Otto Nathan and Heinz Norden, eds., *Einstein on Peace* (New York: Simon & Schuster, 1960), 376.

5. Edward Teller quoted in Gruber, "Manhattan Project Maverick," 87; Edward Teller, "How Dangerous Are Atomic Weapons?," *Bulletin of the Atomic Scientists* 3 (Feb. 1947): 35.

6. Leo Szilard quoted in Smith, *A Peril and a Hope*, 88; Gruber, "Manhattan Project Maverick," 84–87.

7. Smith and Weiner, *Robert Oppenheimer*, 328–32.

8. Memorandum, the President to Dean Acheson, 7 May 1946, Box 201, President's Secretary File, Papers of Harry S Truman; J. Robert Oppenheimer, "Physics in the Contemporary World," *Bulletin of the Atomic Scientists* 4 (Mar. 1948): 66; Ernest Lawrence quoted in Herbert F. York, *Making Weapons, Talking Peace: A Physicist's Odyssey from Hiroshima to Geneva* (New York: Basic Books, 1987), 49.

9. Smith and Weiner, *Robert Oppenheimer*, 315, 317.

10. Robert Oppenheimer quoted in William J. Broad, "The Men Who Made the Sun Rise," *New York Times Book Review*, 8 Feb. 1987, 39; Robert Oppenheimer quoted in Boyer, *By the Bomb's Early Light*, 72.

11. Merle Miller, "The Atomic Scientist in Politics," *Bulletin of the Atomic Scientists* 3 (Sept. 1947): 242; Kevles, *The Physicists*, 351; Marcel C. LaFollette, *Making Science Our Own: Public Images of Science, 1910–1955* (Chicago: The Univ. of Chicago Press, 1990), 51; Boyer, *By the Bomb's Early Light*, 63.

12. Smith, *A Peril and a Hope*, 203–4; Raymond Gram Swing quoted in

Kevles, *The Physicists*, 351; Dexter Masters and Katherine Way, eds., *One World or None* (New York: McGraw-Hill, 1946); Boyer, *By the Bomb's Early Light*, 51, 76–79.

13. Len Ackland, "Fighting for Time," *Chicago Tribune Magazine*, 11 Mar. 1984; Boyer, *By the Bomb's Early Light*, 63–64.

14. Bulletin of the Atomic Scientists, "We Tell the World What Time It Is" (poster); Ackland, "Fighting for Time."

15. Miller, "The Atomic Scientist in Politics," 242, 252; *Harper's* contributor quoted in Kevles, *The Physicists*, 375–76.

16. The image of the Tralfamadorians comes from Martin J. Sherwin, "How Well They Meant," *Bulletin of the Atomic Scientists* 41 (Aug. 1985): 12; Kurt Vonnegut, Jr., *Slaughterhouse-Five* (New York: Dell, 1968), 116–17; LaFollette, 166–67; Smith, *A Peril and a Hope*, 253.

17. Sherwin, *A World Destroyed*, 92; C. P. Snow and Albert Einstein quoted in Rhodes, *The Making of the Atomic Bomb*, 53–54; Robert R. Wilson, "Niels Bohr and the Young Scientists," *Bulletin of the Atomic Scientists* 41 (Aug. 1985): 23.

18. Oppenheimer quoted in Smith, *A Peril and a Hope*, 6.

19. Niels Bohr quoted in Sherwin, *A World Destroyed*, 93; Niels Bohr quoted in Sherwin, "How Well They Meant," 12.

20. Wilson, "Niels Boher," 24–25; Smith, *A Peril and a Hope*, 7–11; Sherwin, *A World Destroyed*, 107–10.

21. *Public Papers of the Presidents of the United States: Harry S Truman, April 12 to December 31, 1945* (Washington, D.C.: U.S. Government Printing Office, 1961), 213; Truman, *Memoirs*, Vol. 1, 532–33.

22. *New York Times*, 24 June 1941; Donovan, *Conflict and Crisis*, 36; Truman, *Memoirs*, Vol. 1, 82.

23. Elting E. Morison, *Turmoil and Tradition: A Study of the Life and Times of Henry L. Stimson* (New York: Atheneum, 1964), 167–68; Diary of Henry L. Stimson, 15 May 1945, Henry L. Stimson Papers; Memorandum, Henry L. Stimson to the President, 11 Sept. 1945, Henry L. Stimson Papers (also located in President's Secretary File, Papers of Harry S Truman; Dean Acheson, *Present at the Creation: My Years in the State Department* (New York: New American Library, 1969), 174; McGeorge Bundy, *Danger and Survival: Choices about the Bomb in the First Fifty Years* (New York: Random House, 1988), 139.

24. Bundy, *Danger and Survival*, 139–41.

25. Ibid., 138; Sherwin, *A World Destroyed*, 315; Diary of Henry L. Stimson, 4 Sept. 1945, Henry L. Stimson Papers.

26. Truman, *Memoirs*, Vol. 1, 527–28; Bundy, *Danger and Survival*, 145.

27. Gallup, *The Gallup Poll*, Vol. 1, 525, 536; Barton J. Bernstein, "The Quest for Security: American Foreign Policy and International Control of Atomic Energy, 1942–1946," *Journal of American History* 60 (Mar. 1974): 1020; Herken, *The Winning Weapon*, 32.

28. Pringle and Spigelman, *The Nuclear Barons*, 42–43; Truman, *Memoirs*, Vol. 1. 533–34; Bundy, *Danger and Survival*, 144–45.

29. Memorandum by the Director of the Office of Scientific Research and Development (Bush) to the Secretary of State, 5 Nov. 1945, *Foreign Relations of the United States: 1945*; Vol. 2, *General: Political and Economic Matters* (Wash-

ington, D.C.: U.S. Government Printing Office, 1967), 69–73; Bundy, *Danger and Survival*, 145–48; Bernstein, "The Quest for Security," 1023–24, 1027–29.

30. Robert Oppenheimer quoted in Barton J. Bernstein and Allen J. Matusow, eds., *The Truman Administration: A Documentary History* (New York: Harper & Row, 1966), 224; David E. Lilienthal, *The Journals of David E. Lilienthal*, Vol. 2, *The Atomic Energy Years, 1945–1950* (New York: Harper & Row, 1964), 10.

31. Arthur Vandenberg quoted in Bernstein, "The Quest for Security," 1028; Memorandum, Frank McNaughton to Don Bermingham, 6 Oct. 1945, Box 8, Papers of Frank McNaughton, Harry S Truman Library.

32. Donovan, *Conflict and Crisis*, 268; Bundy, *Danger and Survival*, 158–59; Bernstein, "The Quest for Security," 1029–30.

33. Pringle and Spigelman, *The Nuclear Barons*, 48–49; Lilienthal, *The Journals of David E. Lilienthal*, Vol. 2, 14–16.

34. Acheson, *Present at the Creation*, 211; Hewlett and Anderson, *The New World*, 536; Bundy, *Danger and Survival*, 159–60; Bernstein, "The Quest for Security," 1030.

35. *A Report on the International Control of Atomic Energy*, Department of State Publication 2498 (Washington, D.C.: U.S. Government Printing Office, 16 Mar. 1946), viii, xii; Bernstein, "The Quest for Security," 1030–32; Bundy, *Danger and Survival*, 160–61.

36. Edward Teller, "The State Dep't Report—'A Ray of Hope,' " *Bulletin of the Atomic Scientists* 1 (1 April 1946): 10; "H.C. Urey on State Dept. Report," *Bulletin of the Atomic Scientists* 1 (1 April 1946): 13; William Higinbotham quoted in Smith, *A Peril and a Hope*, 461.

37. I. F. Stone, "Atomic Pie in the Sky," *The Nation*, 6 Apr. 1946, 391; Robert Frost, *Complete Poems of Robert Frost, 1949* (New York: Henry Holt, 1949), 569.

38. Bundy, *Danger and Survival*, 161–62; Bernstein, "The Quest for Security," 1032–33; Lilienthal, *The Journals of David E. Lilienthal*, Vol. 2, 30.

39. Bernstein, "The Quest for Security," 1032–33; List of the President's Appointments, 16 Mar. 1946, Box 83, President's Secretary File, Papers of Harry S Truman; James Byrnes quoted in Lilienthal, *The Journals of David E. Lilienthal*, Vol. 2, 59.

40. Harry Truman quoted in Herken, *The Winning Weapon*, 160; Bernard Baruch quoted in Bernstein, "The Quest for Security," 1033–34.

41. Lilienthal, *The Journals of David E. Lilienthal*, Vol. 2, 32; Bernard Baruch quoted in Herken, *The Winning Weapon*, 161.

42. Bundy, *Danger and Survival*, 163–65; George T. Mazuzan, "American Nuclear Policy," in John M. Carroll and George C. Herring, eds., *Modern American Diplomacy* (Wilmington, Del.: Scholarly Resources, 1986), 150.

43. Bernard Baruch quoted in Smith, *A Peril and a Hope*, 475; Bundy, *Danger and Survival*, 165–66; Bernstein, "The Quest for Security," 1037.

44. Pringle and Spigelman, *The Nuclear Barons*, 54; Hanson Baldwin, Anne O'Hare McCormick, and the Hearst press quoted in Hewlett and Anderson, *The New World*, 582; Smith, *A Peril and a Hope*, 476.

45. Smith, *A Peril and a Hope*, 476; Bernstein, "The Quest for Security," 1038; Hewlett and Anderson, *The New World*, 583.

46. Mazuzan, "American Nuclear Policy," 150; Bernstein, "The Quest for Security," 1038–39; Bundy, *Danger and Survival*, 166.

47. Bundy, *Danger and Survival*, 166; Bernstein, "The Quest for Security," 1042–44.

48. Kevles, *The Physicists*, 339–40; Bernstein, "The Quest for Security," 1044.

49. George T. Mazuzan and J. Samuel Walker, *Controlling the Atom: The Beginnings of Nuclear Regulation, 1946–1962* (Berkeley, Calif.: Univ. of California Press, 1984), 3; Donovan, *Conflict and Crisis*, 133; Smith, *A Peril and a Hope*, 129–30; Kevles, *The Physicists*, 349.

50. Smith, *A Peril and a Hope*, 129–31; Herbert Anderson quoted in Kevles, *The Physicists*; 351.

51. Memorandum, Frank McNaughton to Don Bermingham, 6 Oct. 1945, Papers of Frank McNaughton; Donovan, *Conflict and Crisis*, 133.

52. Kevles, *The Physicists*, 350; Donovan, *Conflict and Crisis*, 133–34.

53. Harry S Truman, *Memoirs*, Vol. 2, *Years of Trial and Hope* (Garden City, N.Y.: Doubleday, 1956), 3.

54. Kevles, *The Physicists*, 351; James Newman quoted in Smith, *A Peril and a Hope*, 262.

55. Donovan, *Conflict and Crisis*, 207; George T. Mazuzan and Roger R. Trask, "An Outline History of Nuclear Regulation and Licensing, 1946–1979," Apr. 1979, 1–8, in Historical Office, Office of the Secretary, Nuclear Regulatory Commission, Washington, D.C.; Mazuzan and Walker, *Controlling the Atom*, 3–4.

56. Mazuzan and Walker, *Controlling the Atom*, 4–5; Donovan, *Conflict and Crisis*, 272–274; Ferrell, *Off the Record*, 113.

3. Strategy, Weaponry, and the Early Arms Race

1. Scientists' letter quoted in Tom Wicker, "That Bang Resounds to This Day," *International Herald-Tribune*, 19 July 1985; Freeman Dyson quoted in David Quammen, "The Conscience of the Young Scientist," *Esquire*, Dec. 1985, 223.

2. Alexander Sachs to Henry Wallace, 6 Sept. 1945, Box 199, President's Secretary File, Papers of Harry S Truman; Winston Churchill speech, 1 Mar. 1955, *Parliamentary Debates (Hansard), Fifth Series—Volume 537, House of Commons, Official Report, Session 1954–55* (London: Her Majesty's Stationery Office, 1955), 1900; Bundy, *Danger and Survival*, 198.

3. Truman, *Memoirs*, Vol. 1, 87; Leo Szilard, "Reminiscences," in Donald Fleming and Bernard Bailyn, eds., *Perspectives in American History*, Vol. 2, *1968* (Cambridge, Mass.: Charles Warren Center for Studies in American History, 1968), 128.

4. Curtis LeMay quoted in John Newhouse, *War and Peace in the Nuclear Age* (New York: Alfred A. Knopf, 1989), 70; Herken, *The Winning Weapon*, 211, 214; Air Force report quoted in Aaron L. Friedberg, "A History of the U.S. Strategic 'Doctrine'—1945 to 1980," *Journal of Strategic Studies* 3 (Dec. 1980): 40; Memorandum by the Commanding General, Manhattan Engineer District (Groves), 2 Jan. 1946, *Foreign Relations of the United States: 1946*, Vol. 1, *General; The United Nations* (Washington, D.C.: U.S. Government Printing Office, 1972), 1203.

5. John L. Harper, "Henry Stimson and the Origin of America's Attachment to Atomic Weapons," *SAIS Review* 5 (Summer–Fall 1985): 26.

6. Vyacheslav Molotov and James Byrnes quoted in Herken, *The Winning Weapon*, 48.

7. Barton J. Bernstein, "The Challenges and Dangers of Nuclear Weapons: American Foreign Policy and Strategy, 1941–1961," *Foreign Service Journal*, Sept. 1978, 11; Herken, *The Winning Weapon*, 216; Herbert Y. Schandler, "U.S. Policy on the Use of Nuclear Weapons, 1945–1975," Congressional Research Service, 1975, 6.

8. Stockpile Figures, U.S. Department of Energy Archives, Germantown, Md.; David Alan Rosenberg, "The Origins of Overkill: Nuclear Weapons and American Strategy, 1945–1960," *International Security* 7 (Spring 1983): 14–15; David Alan Rosenberg, "American Atomic Strategy and the Hydrogen Bomb Decision," *Journal of American History* 66 (June 1979): 65–66.

9. Rosenberg, "The Origins of Overkill," 11; Gregg Herken, *Counsels of War* (New York: Alfred A. Knopf, 1985), 27.

10. Karl von Clausewitz quoted in Bernard Brodie, "Strategy," in David L. Sills, ed., *International Encyclopedia of the Social Sciences*, Vol. 15 (New York: Macmillan and Free Press, 1968), 284; Bernard and Fawn Brodie, *From Crossbow to H-Bomb*, Revised and Enlarged Edition (Bloomington, Ind.: Indiana Univ. Press, 1973), 264.

11. Bernard Brodie quoted in Herken, *Counsels of War*, 8.

12. Bernard Brodie, "War in the Atomic Age," 24, 28, and Bernard Brodie, "Implications for Military Strategy," in Bernard Brodie, ed., *The Absolute Weapon: Atomic Power and World Order* (New York: Harcourt Brace, 1946), 75–76; Herken, *Counsels of War*, 9–10, 14.

13. Herken, *Counsels of War*, 26–27; Herken, *The Winning Weapon*, 214–15; Rosenberg, "American Atomic Strategy," 71.

14. The Evaluation of the Atomic Bomb as a Military Weapon: The Final Report of the Joint Chiefs of Staff Evaluation Board for Operation Crossroads, 30 June 1947, Box 202, President's Secretary File, Papers of Harry S Truman; Final Report also quoted in Herken, *The Winning Weapon*, 226; Rosenberg, "American Atomic Strategy," 66–67.

15. Herken, *The Winning Weapon*, 27, 33; Rosenberg, "The Origins of Overkill," 22–23; Herbert York, "Vertical Proliferation," *Bulletin of the Atomic Scientists* 38 (Dec. 1982): 49; Bundy, *Danger and Survival*, 203.

16. Rosenberg, "The Origins of Overkill," 19–20; Richard Smoke, *National Security and the Nuclear Dilemma: An Introduction to the American Experience* (Reading, Mass.: Addison-Wesley, 1984), 53; the Harvard Nuclear Study Group

(Albert Carnesale, Paul Doty, Stanley Hoffmann, Samuel P. Huntington, Joseph S. Nye, Jr., Scott D. Sagan), *Living with Nuclear Weapons* (New York: Bantam, 1983), 79.

17. Rosenberg, "American Atomic Strategy," 68–69; Herken, *The Winning Weapon*, 264, 268–70; Memorandum by the Director of the Office of Far Eastern Affairs (Butterworth), 15 Sept. 1948, *Foreign Relations of the United States: 1948*, Vol. 1, *General; The United Nations* (Washington, D.C.: U.S. Government Printing Office, 1976), 630–31; Walter Millis, ed., *The Forrestal Diaries* (New York: Viking, 1951), 487; Lilienthal, *The Journals of David E. Lilienthal*, Vol. 2, 271.

18. Rosenberg, "The Origins of Overkill," 15.

19. Harmon report quoted in Rosenberg, "American Atomic Strategy," 72–73.

20. Bernard Brodie quoted in Herken, *Counsels of War*, 28, and in Rosenberg, "The Origins of Overkill," 18; Fred Kaplan, "Strategic Thinkers," *Bulletin of the Atomic Scientists* 38 (Dec. 1982): 51.

21. I. I. Rabi quoted in U.S. Atomic Energy Commission, *In the Matter of J. Robert Oppenheimer, Transcript of Hearing before Personnel Security Board*, 467; Record of the Meeting of the Joint Congressional Committee on Atomic Energy, 20 July 1949, *Foreign Relations of the United States: 1949*, Vol. 1, *National Security Affairs, Foreign Economic Policy* (Washington, D.C.: U.S. Government Printing Office, 1976), 491; Interdepartmental Intelligence Study, *Foreign Relations of the United States: 1947*, Vol. 1, *General; The United Nations* (Washington, D.C.: U.S. Government Printing Office, 1973), 905; Pringle and Spigelman, *The Nuclear Barons*, 42–43; conversation between Truman and Oppenheimer quoted in Davis, *Lawrence & Oppenheimer*, 260.

22. Richard G. Hewlett and Francis Duncan, *Atomic Shield, 1947–1952*, Vol. 2, *A History of the United States Atomic Energy Commission* (Washington, D.C.: U.S. Atomic Energy Commission, 1972), 362–68; *New York Times*, 24 Sept. 1949.

23. Hewlett and Duncan, *Atomic Shield*, 368; Harold C. Urey quoted in Eric F. Goldman, *The Crucial Decade—and After: America, 1945–1960* (New York: Vintage, 1960), 100; Ackland, "Fighting for Time."

24. Edward Teller, "The Work of Many People," *Science* 121 (25 Feb. 1955): 268; Enrico Fermi quoted in Herbert F. York, *The Advisors: Oppenheimer, Teller, and the Superbomb* (Stanford, Calif.: Stanford Univ. Press, 1976), 21.

25. York, *The Advisors*, 4; Rhodes, *The Making of the Atomic Bomb*, 189–90, 192–94, 538; "Hydrogen Bomb Deviser: Dr. Edward Teller," *New York Times*, 27 June 1957; Edward Teller, "To Sagittarius," *Bulletin of the Atomic Scientists* 7 (Jan. 1951): 22.

26. Bundy, *Danger and Survival*, 205; Edward Teller quoted in Herken, *Counsels of War*, 57.

27. Teller, "How Dangerous Are Atomic Weapons?," 35–36.

28. York, *The Advisors*, 63; Hewlett and Duncan, *Atomic Shield*, 369; Herken, *Counsels of War*, 57.

29. Bundy, *Danger and Survival*, 205–6; York, *The Advisors*, 63–65; Herbert F. York, "The Debate over the Hydrogen Bomb," *Scientific American* 233 (Oct. 1975): 106–7.

30. Samuel Goudsmit quoted in Daniel Lang, *From Hiroshima to the Moon: Chronicles of Life in the Atomic Age* (New York: Simon & Schuster, 1959), 219; Herken, *Counsels of War*, 55; Rosenberg, "American Atomic Strategy," 80; York, "The Debate over the Hydrogen Bomb," 108.

31. York, *The Advisors*, 42, 44; *Washington Post*, 18 Nov. 1949; *New York Times*, 17 Jan. 1950.

32. U.S. Atomic Energy Commission, Historical Document Number 349, Appendix I, in York, *The Advisors*, 155–62.

33. York, *The Advisors*, 57; Lilienthal, *The Journals of David E. Lilienthal*, Vol. 2, 594; The Chairman of the U.S. Atomic Energy Commission (Lilienthal) to President Truman, 9 Nov. 1949, *Foreign Relations of the United States: 1949*, Vol. 1, 582.

34. Winkler, "Harry S Truman," 79.

35. Bundy, *Danger and Survival*, 211; The Chairman of the Joint Committee on Atomic Energy (McMahon) to President Truman, 21 Nov. 1949, *Foreign Relations of the United States: 1949*, Vol. 1, 588–95; Lewis L. Strauss, *Men and Decisions* (Garden City, N.Y.: Doubleday, 1962), 219–22, 440; Rosenberg, "American Atomic Strategy," 81–82.

36. Memorandum of Telephone Conversation, by the Secretary of State, 19 Jan. 1950, *Foreign Relations of the United States: 1950*, Vol. 1, *National Security Affairs; Foreign Economic Policy* (Washington, D.C.: U.S. Government Printing Office, 1977), 511; Bundy, *Danger and Survival*, 212.

37. Winkler, "Harry S Truman," 79–80; Bundy, *Danger and Survival*, 213; Lilienthal, *The Journals of David E. Lilienthal*, Vol. 2, 633; Report by the Special Committee of the National Security Council to the President, 31 Jan. 1950, *Foreign Relations of the United States: 1950*, Vol. 1, 513; *New York Times*, 1 Feb. 1950.

38. Winkler, "Harry S Truman," 80–81; Eben Ayers quoted in Herken, *The Winning Weapon*, 320–21.

39. Albert Einstein, Hans Bethe and the physicists, and Robert Oppenheimer all quoted in Philip M. Stern, *The Oppenheimer Case: Security on Trial* (New York: Harper & Row, 1969), 154–55.

40. Harry Truman quoted in Margaret Truman, *Harry S Truman*, 418–19; Herken, *The Winning Weapon*, 321.

41. Hans A. Bethe, "Observations on the Development of the H-Bomb," Appendix II, in York, *The Advisors*, 168–69; William J. Broad, "Rewriting the History of the H-Bomb," *Science* 218 (19 Nov. 1982): 769–70; Daniel Hirsch and William G. Mathews, "The H-Bomb: Who Really Gave Away the Secret?," *Bulletin of the Atomic Scientists* 46 (Jan.–Feb. 1990): 24–25

42. Bethe, "Observations," 171–72; Hirsch and Mathews, "The H-Bomb," 26–27; Broad, "Rewriting the History of the H-Bomb," 770; Hans Bethe quoted in Rhodes, *The Making of the Atomic Bomb*, 773; Robert Oppenheimer quoted in Jungk, *Brighter Than a Thousand Suns*, 296.

43. York, "The Debate over the Hydrogen Bomb," 111; York, *The Advisors*, 82; Newhouse, *War and Peace in the Nuclear Age*, 80; Luis Alvarez quoted in Herken, *Counsels of War*, 59.

44. York, "The Debate over the Hydrogen Bomb," 111.

45. Bulletin of the Atomic Scientists, "We Tell the World What Time It Is" (poster); Val Peterson to Lewis Strauss, 3 May 1954, Box 71, Records of the Federal Civil Defense Administration (FCDA), in Records of the Office of Civil and Defense Mobilization, Washington National Records Center, Suitland, Md.; Bertrand Russell, "The Danger to Mankind," *Bulletin of the Atomic Scientists* 10 (Jan. 1954): 8–9; Winston Churchill speech, 1 Mar. 1955, in *Parliamentary Debates (Hansard)*, 1894–95; Bundy, *Danger and Survival*, 198.

46. J. Robert Oppenheimer, "Atomic Weapons and American Policy," *Foreign Affairs* 31 (July 1953): 529.

47. York, "The Debate over the Hydrogen Bomb," 113; McGeorge Bundy, "The Missed Chance to Stop the H-Bomb," *New York Review of Books* 29 (13 May 1982), 13, 16, 20; Newhouse, *War and Peace in the Nuclear Age*, 80; Vannevar Bush quoted in Hirsch and Mathews, "The H-Bomb," 30.

48. U.S. Atomic Energy Commission, *In the Matter of J. Robert Oppenheimer: Transcript of Hearing*, 837; Bundy, *Danger and Survival*, 305; Stern, *The Oppenheimer Case*, 214–18.

49. Mazuzan and Walker, *Controlling the Atom*, 5–6; Bundy, *Danger and Survival*, 206, 306; Stanley A. Blumberg and Gwinn Owens, *Energy and Conflict: The Life and Times of Edward Teller* (New York: G.P. Putnam's Sons, 1976), 198.

50. Stern, *The Oppenheimer Case*, 129–30; Richard Pfau, *No Sacrifice Too Great: The Life of Lewis L. Strauss* (Charlottesville, Va.: Univ. Press of Virginia, 1984), 159, 162, 184; Bundy, *Danger and Survival*, 306–8.

51. Herken, *Counsels of War*, 69–70; Kevles, *The Physicists*, 380; U.S. Atomic Energy Commission, *In the Matter of J. Robert Oppenheimer: Transcript of Hearing*, 710, 726; Stern, *The Oppenheimer Case*, 334–40.

52. U.S. Atomic Energy Commission, *In the Matter of J. Robert Oppenheimer: Texts of Principal Documents and Letters of Personnel Security Board, General Manager, Commissioners* (Washington, D.C.: U.S. Government Printing Office, 1954), 13, 43–48, 54; Bundy, *Danger and Survival*, 310–11.

53. William Borden quoted in Herken, *Counsels of War*, 72; Brian Balogh, *Chain Reaction: Expert Debate & Public Participation in American Commercial Nuclear Power, 1945–1975* (New York: Cambridge Univ. Press, 1991), 58–59.

54. NSC 162/2, 30 Oct. 1953, paragraph 39.B, *Foreign Relations of the United States: 1952–1954*, Vol. 2, *National Security Affairs* (Washington, D.C.: U.S. Government Printing Office, 1984), 593; Dwight D. Eisenhower, "Annual Message to the Congress on the State of the Union," 7 Jan. 1954, *Public Papers of the Presidents of the United States: Dwight D. Eisenhower, 1954* (Washington, D.C.: U.S. Government Printing Office, 1960), 10; John Foster Dulles, "The Evolution of Foreign Policy," *Department of State Bulletin* 30 (25 Jan. 1954): 108; Samuel F. Wells, Jr., "The Origins of Massive Retaliation," *Political Science Quarterly* 96 (Spring 1981): 33–34; Allan M. Winkler, "The Nuclear Question," in Robert H. Bremner, Gary W. Reichard, and Richard J. Hopkins, eds., *American Choices: Social Dilemmas and Public Policy since 1960* (Columbus, Ohio: Ohio State Univ. Press, 1986), 135–36.

55. Bundy, *Danger and Survival*, 319.

56. Bernstein, "Challenges and Dangers," 14–15; Bundy, *Danger and Sur-*

vival, 238–44, 268–70; Schandler, "U.S. Policy on the Use of Nuclear Weapons," 11–12.

57. David Alan Rosenberg, " 'A Smoking Radiating Ruin at the End of Two Hours': Documents on American Plans for Nuclear War with the Soviet Union, 1954–1955," *International Security* 6 (Winter 1981–82): 11; Friedberg, "A History of the U.S. Strategic 'Doctrine,' " 42, 47; Rosenberg, "The Origins of Overkill," 4, 7.

58. Kaplan, "Strategic Thinkers," 51–52.

59. Henry A. Kissinger, *Nuclear Weapons and Foreign Policy* (New York: Harper & Brothers, 1957), 20, 172–74; Herman Kahn, *On Thermonuclear War,* Second Edition with Index (Princeton, N.J.: Princeton Univ. Press, 1961), 22; Herken, *Counsels of War,* 99, 206–7; James Newman, "Books: Two Discussions of Thermonuclear War," *Scientific American* 204 (Mar. 1961): 197.

60. Dwight D. Eisenhower quoted in Rosenberg, "The Origins of Overkill," 8.

4. Fear of Fallout

1. Ralph E. Lapp, "Civil Defense Faces New Peril," *Bulletin of the Atomic Scientists* 10 (Nov. 1954): 350; Ralph E. Lapp, *The Weapons Culture* (New York: W.W. Norton, 1968), 138, 140.

2. Catherine Caufield, *Multiple Exposures: Chronicles of the Radiation Age* (New York, Harper & Row, 1989), 3–6, 22–23; Stephen Hilgartner, Richard C. Bell, Rory O'Connor, *Nukespeak: The Selling of Nuclear Technology in America* (New York: Penguin, 1983), 2–4; Weart, *Nuclear Fear,* 5; Rhodes, *The Making of the Atomic Bomb,* 42; Gerard H. Clarfield and William M. Wiecek, *Nuclear America: Military and Civilian Nuclear Power in the United States, 1940–1980* (New York: Harper & Row, 1984), 8–9; Yvonne Baskin, "Too Intimate With the Atom," *New York Times Book Review,* 7 Jan. 1990, 32.

3. *Journal of the American Medical Association* and E. P. Davis quoted in J. Samuel Walker, "The Controversy Over Radiation Safety: A Historical Overview," *JAMA—Journal of the American Medical Association* 262 (4 Aug. 1989): 664; *Photography* quoted in Caufield, *Multiple Exposures,* 7; Hilgartner, Bell, O'Connor, *Nukespeak,* 3.

4. Sir William Ramsay and the *New York Times* quoted in Hilgartner, Bell, O'Connor, *Nukespeak,* 4–5.

5. Caufield, *Multiple Exposures,* 8–9, 13.

6. John Hall Edwards quoted in Hilgartner, Bell, O'Connor, *Nukespeak,* 8.

7. Caufield, *Multiple Exposures,* 9–10.

8. Caufield, *Multiple Exposures,* 29–32; Barton C. Hacker, *The Dragon's Tail: Radiation Safety in the Manhattan Project, 1942–1946* (Berkeley and Los Angeles: Univ. of California Press, 1987), 20–23.

9. Mazuzan and Walker, *Controlling the Atom,* 34; Walker, "The Controversy Over Radiation Safety," 665; Background Information on National Com-

mittee on Radiation Protection and Measurement, Prepared by the Public Health Service, Department of Health, Education and Welfare, Mar. 1959, RG 326, U.S. Atomic Energy Commission, Secretariat Collection, 4928, MH & S 3 Radiation, U.S. Department of Energy Archives.

10. Caufield, *Multiple Exposures*, 18–21; Lauriston S. Taylor, "The Origin and Significance of Radiation Dose Limits for the Population," WASH–1336, presented at the AEC Scientific and Technical Symposium, 17 Aug. 1973, Document 14654, Records of the Department of Energy, Coordination and Information Center, Reynolds Electrical & Engineering Co., Inc., Las Vegas, Nev. Hacker, *The Dragon's Tail*, 14–19; Walker, "The Controversy Over Radiation Safety," 665.

11. Walker, "The Controversy Over Radiation Safety," 665; Hacker, *The Dragon's Tail*, 67–68, 73; Caufield, *Multiple Exposures*, 59–63; Clifford T. Honicker, "The Hidden Files: America's Radiation Victims," *New York Times Magazine*, 19 Nov. 1989, 39–40.

12. Hacker, *The Dragon's Tail*, 158; Clarfield and Wiecek, *Nuclear America*, 202; J. Samuel Walker, "Writing the History of Nuclear Energy: The State of the Art," *Diplomatic History* 9 (Fall 1985): 381; J. Samuel Walker, *Containing the Atom: Nuclear Regulation in a Changing Environment, 1963–1971* (Berkeley, Calif.: Univ. of California Press, 1992), 298–99; Lauriston Taylor quoted in Caufield, *Multiple Exposures*, 120.

13. Jonathon M. Weisgall, "Micronesia and the Nuclear Pacific since Hiroshima," *SAIS Review* 5 (Summer–Fall 1985): 43–44; Hilgartner, Bell, O'Connor, *Nukespeak*, 72–73; W.H.P. Blandy, "Report on Bikini," *13 ND Naval Reserve Bulletin*, 1 Nov. 1946, Archives, Univ. of Oregon, Eugene, Oreg.; Goulden, *The Best Years*, 263.

14. Caufield, *Multiple Exposures*, 93–94; David Bradley, *No Place to Hide, 1946/1984* (Hanover, N.H.: Univ. Press of New England, 1983), xvii, 103–4, 125, 165; Paul Boyer, "Physicians Confront the Apocalypse: The American Medical Profession and the Threat of Nuclear War," *JAMA—Journal of the American Medical Association* 254 (2 Aug. 1985): 634–35; Boyer, *By the Bomb's Early Light*, 91.

15. Atomic Energy Commission, Selection of a Continental Atomic Test Site: Report by the Director of Military Application, AEC 141/7, 13 Dec. 1950, Energy History Collection, U.S. Department of Energy Archives; Philip G. Schrag, "Seeing Ground Zero in Nevada," *New York Times*, 12 Mar. 1989; Caufield, *Multiple Exposures*, 103–4; Howard Ball, *Justice Downwind: America's Atomic Testing Program in the 1950's* (New York: Oxford Univ. Press, 1986), 59–83.

16. Schrag, "Seeing Ground Zero"; "Atomic Blast at Yucca Flat," *Business Week*, 3 May 1952, 110–12; Chet Huntley quoted in Weart, *Nuclear Fear*, 132; Leonard Slater: "Atomic Tests: The World Blew Up," *Newsweek*, 19 Feb. 1951, 25; "Testimony Links Deaths to 'Beautiful Mushroom,' " *Eugene* [Oregon] *Register-Guard*, 29 Sept. 1982; Hales, "The Atomic Sublime."

17. H. McK. Roper (Military Liaison Committee) to the Atomic Energy Commission, 16 July 1951, Energy History Collection, U.S. Department of Energy Archives; Howard L. Rosenberg, *Atomic Soldiers: American Victims of Nuclear*

Experiments (Boston: Beacon Press, 1980), 40–41; Caufield, *Multiple Exposures*, 107–8; Clarfield and Wiecek, *Nuclear America*, 203–5; Harold Kinne quoted in "Greasewood Fires and Man's Most Terrible Weapon," *Newsweek*, 30 Mar. 1953, 31; Blanche Wiesen Cook, "The Legacy of Atomic Testing," *New York Times Book Review*, 1 Aug. 1982, 3; *Las Vegas Sun*, 21 Apr. 1952, 23 Apr. 1952, 25 Mar. 1953; *New York Times*, 17 Mar. 1953.

18. *Los Angeles Examiner*, 17 Apr. 1955; Lyle Borst quoted in Caufield, *Multiple Exposures*, 109.

19. Caufield, *Multiple Exposures*, 110–11; Philip L. Fradkin, *Fallout: An American Nuclear Tragedy* (Tucson, Ariz.: The Univ. of Arizona Press, 1989), 1–26; *New York Times*, 20 May 1953, 19 Sept. 1982; *Salt Lake City Tribune*, 20 May 1953; *San Francisco Examiner*, 20 May 1953; AEC press release—Study Shows Radioactivity from Atomic Tests Did Not Cause Sheep Deaths, 13 Jan. 1954, Document 14045, Records of the Department of Energy, Coordination and Information Center; Lewis L. Strauss to Douglas R. Stringfellow, 9 Mar. 1954, RG 326, U.S. Atomic Energy Commission, Secretariat Collection, 4928, MH & S 3 Radiation, U.S. Department of Energy Archives; Edward Diamond, Memorandum on *Bulloch et al.* v. *United States*, 30 Nov. 1956, Energy History Collection, U.S. Department of Energy Archives; R. Jeffrey Smith, "Scientists Implicated in Atom Test Deception," *Science* 218 (5 Nov. 1982): 545.

20. Caufield, *Multiple Exposures*, 113.

21. *Fallout* (pamphlet), 20–21, Box 20, Records of the Committee for a Sane Nuclear Policy, Swarthmore College Library, Swarthmore, Pa.; "Radiation: 'Glimpse Into Hell,' " *Newsweek*, 13 Jan. 1958, 78; Ralph E. Lapp, *The Voyage of the* Lucky Dragon (New York: Harper & Brothers, 1958); Richard Hudson and Ben Shahn, *Kuboyama and the Saga of the* Lucky Dragon (New York: Thomas Yoseloff, 1965); Lewis Strauss quoted in Robert A. Divine, *Blowing on the Wind: The Nuclear Test Ban Debate, 1954–1960* (New York: Oxford Univ. Press, 1978), 11; Statement by Ambassador Allison on *Fukuryu Maru* Accident, 9 Apr. 1954, Box 73, Papers of Sidney R. Yates, Harry S Truman Library; Lapp, "Civil Defense Faces New Peril," 351; Caufield, *Multiple Exposures*, 114.

22. Charles E. Wilson quoted in Daniel S. Greenberg, *The Politics of Pure Science* (New York: New American Library, 1967), 273; Eugene Rabinowitch quoted in Donald W. Cox, *America's New Policy Makers: The Scientists' Rise to Power* (New York: Chilton, 1964), 39; Kevles, *The Physicists*, 382–84.

23. Herken, *Counsels of War*, 177; Mazuzan and Walker, *Controlling the Atom*, 44; Caufield, *Multiple Exposures*, 127.

24. A. H. Sturtevant and Edwin C. Johnson quoted in Carolyn Kopp, "The Origins of the American Scientific Debate over Fallout Hazards," *Social Studies of Science* 9 (1979): 405–13; Caufield, *Multiple Exposures*, 127–28.

25. Kopp, "Debate over Fallout Hazards," 413–15.

26. Steven M. Spencer, "Fallout: The Silent Killer, Part One," *The Saturday Evening Post*, 29 Aug. 1959, 89; Kopp, "Debate over Fallout Hazards," 416, 418; Caufield, *Multiple Exposures*, 128–29; Linus Pauling, *Every Test Kills* (pamphlet), Box B–14, Records of the National Committee for a Sane Nuclear Policy; "The

Atom: Political Fallout," *Newsweek*, 17 June 1957, 38; Graham DuShane, "Loaded Dice," *Science* 125 (17 May 1957): 963; E. B. Lewis, "Leukemia and Ionizing Radiation," *Science* 125 (17 May 1957): 965–72.

27. Poul Anderson and F. N. Waldrop, "Tomorrow's Children," and Edward Grendon, "The Figure," in Groff Conklin, ed., *A Treasury of Science Fiction* (New York: Bonanza, 1980), 28–29, 32–34, 38–39, 70–71.

28. Judith Merril, "That Only a Mother," in Pamela Sargent, ed., *Women of Wonder: Science Fiction Stories by Women about Women* (New York: Penguin, 1978), 57–58, 66

29. Boyer, *By the Bomb's Early Light, 354; Weart, Nuclear Fear, 191.*

30. *Weart, Nuclear Fear*, 191; Stan Lee and J. Kirby, "The Hulk," in *Marvel Tales Annual, #1, 1964* (New York: Non-Pareil Publishing Corp., 1964); Stan Lee, *Origins of Marvel Comics* (New York: Simon & Schuster, 1974), 75–93. Allan M. Winkler, "American Hopes and Fears in the Nuclear Age," in *Sweet Reason: A Journal of Ideas, History & Culture*, published by the Oregon Committee for the Humanities, Portland, Oreg. 5 (Fall 1986): 33–34.

31. Stan Lee and S. Ditko, "Spider-Man!" in *Marvel Tales Annual, #1, 1964*.

32. Winkler, "American Hopes and Fears in the Nuclear Age," 34.

33. *L'il Abner*, script property of Tams-Witmark Music Library, Inc., New York, 1/4/27–28.

34. Tom Lehrer, *Too Many Songs by Tom Lehrer* (New York: Pantheon, 1981), 20–23, 81–87; 146–47, 153; "Songs by Tom Lehrer," Number 6216, Reprise Records, n.d.; "An Evening Wasted With Tom Lehrer," Number 6199, Reprise Records, n.d.

35. *The Joan Baez Songbook* (New York: Ryerson Music Publishers, 1964), 170–73.

36. Nevil Shute, *On the Beach* (New York: William Morrow, 1957), 12; Joseph Keyerleber, "On the Beach," in Jack G. Shaheen, ed., *Nuclear War Films* (Carbondale and Edwardsville, Ill.: Southern Illinois Univ. Press, 1978), 31–38; Winkler, "American Hopes and Fears in the Nuclear Age," 36.

37. "How Fatal Is the Fall-Out?" *Time*, 22 Nov. 1954, 79; Norman Isaac Silber, *Test and Protest: The Influence of Consumers Union* (New York: Holmes & Meier, 1983), 104; George H. Gallup, *The Gallup Poll. Public Opinion, 1935–1971*, Vol. 2, *1949–1958* (New York: Random House, 1972), 1,322, 1,488.

38. Sidney Kraus, Reuben Mehling, and Elaine El-Assal, "Mass Media and the Fallout Controversy," *Public Opinion Quarterly* 27 (Summer 1963): 205; "The Contaminators: A Statement by the Editors of *Playboy*," in Fallout-Nuclear Testing, 1959 folder, Box 18, Papers of Sidney R. Yates; Spencer, "Fallout," 27.

39. Clarfield and Wiecek, *Nuclear America*, 214; Walker, "The Controversy Over Radiation Safety," 666; *New York Times*, 7 Apr. 1963.

40. "The Milk We Drink," *Consumer Reports* 24 (Mar. 1959): cover, 102–11; Silber, *Test and Protest*, 103–110.

41. Silber, *Test and Protest*, 110–12.

42. *Thirteenth Semiannual Report of the Atomic Energy Commission* (Washington, D.C.: U.S. Government Printing Office, 1953), 125; Lyle B. Borst to

Gordon Dean, 16 May 1953, and John D. Bugher to Lyle B. Borst, 1 July 1953, RG 326, U.S. Atomic Energy Commission, Secretariat Collection, Box 1258, MH & S 3 Radiation, U.S. Department of Energy Archives.

43. U.S. Atomic Energy Commission, *Atomic Test Effects in the Nevada Test Site Region*, Jan. 1955, 1, 15–20.

44. Atomic Energy Commission, Minutes of Meeting No. 1065, 10 Mar. 1955, Document 877, Records of the Department of Energy, Coordination and Information Center; Clarfield and Wiecek, *Nuclear America*, 221; Mazuzan and Walker, *Controlling the Atom*, 51–52; "The Philosophers' Stone," *Time*, 15 Aug. 1955, 46–48, 50; Remarks Prepared by Dr. Willard F. Libby for Delivery before the University of New Hampshire Distinguished Lecture Series, 11 Apr. 1957, Document 12650, and Remarks Prepared by Dr. Willard F. Libby for Delivery at Amherst College, 30 Apr. 1958, Document 5046, Records of the Department of Energy, Coordination and Information Center; Willard F. Libby quoted in "Fallout Cover-Up," *Newsweek*, 30 Apr. 1979, 26.

45. Federal Civil Defense Administration, *Facts about Fallout*, n.d., and Federal Civil Defense Administration, *What You Should Know about . . . Radioactive Fallout*, n.d., in Box B–8, Records of the Committee for a Sane Nuclear Policy; Dean Pohlenz to Governor Peterson, 23 Sept. 1955, Box 89, Records of the Federal Civil Defense Administration; Robert Griffith, "Dwight D. Eisenhower and the Corporate Commonwealth," *American Historical Review* 87 (Feb. 1982): 93–94, 111; Gordon Dean quoted in Howard L. Rosenberg, *Atomic Soldiers*, 65; Mazuzan and Walker, *Controlling the Atom*, 258; Caufield, *Multiple Exposures*, 130; Minutes of Cabinet Meeting, 6 Nov. 1959, Box 14; Cabinet Agenda for Friday, 11 Dec. 1959, Box 15; Minutes of Cabinet Meeting, 11 Dec. 1959, Box 15; Morse Salisbury to Richard Hirsch, 11 Dec. 1959, Box 15; Possible Questions and Suggested Answers on the Film "On the Beach," 8 Dec. 1959, Box 15; U.S. Information Agency Message, 4 Dec. 1959—all above documents in Whitman Cabinet Series, Papers of Dwight D. Eisenhower, Dwight D. Eisenhower Library, Abilene, Kans.

46. Robert E. Sherwood, "Please Don't Frighten Us," *Atlantic Monthly*, Feb. 1949, 77; Michael Amrine, "The Issue of Fall-Out," *Current History*, Oct. 1957, 226.

47. "A Short History of SANE" (typescript) n.d., Box 1, Folder: Histories of SANE; and "SANE Is Ten: 1957–1967" (typescript), n.d., Box 5, Folder: Speeches & Articles—Donald Keys—both documents in Records of the National Committee for a Sane Nuclear Policy; *New York Times*, 15 Nov. 1957; Lawrence S. Wittner, *Rebels against War: The American Peace Movement, 1941–1960* (New York: Columbia Univ. Press, 1969), 242–45.

48. *The Effects of Nuclear War*, a report from the National Committee for a Sane Nuclear Policy, n.d., Box 10, Folder: Literature, 1957–1962; sketches for "This is the Bomb that Jack Built," Box B–15; Ed (————) to Trevor Thomas, 16 May 1958, Box B–15—all documents in Records of the National Committee for a Sane Nuclear Policy; *New York Times*, 16 Apr. 1962; Godfrey Hodgson, *America in Our Time* (Garden City, N.Y.: Doubleday, 1976), 277.

49. Amy Swerdlow, "Ladies Day at the Capitol: Women Strike for Peace Versus HUAC," *Feminist Studies* 8 (Fall 1982): 493–520; Sara Evans, *Born for*

Liberty: A History of Women in America (New York: Free Press, 1989), 263–64, 268.

50. Divine, *Blowing on the Wind*, 72, 86–102, 143, 147–49, 186–87, 200, 238–39.

5. Civil Defense

1. This entire chapter draws extensively on Allan M. Winkler, "A 40-Year History of Civil Defense," *Bulletin of the Atomic Scientists* 40 (June–July 1984): 16–22.

2. Elwyn A. Mauck, "History of Civil Defense in the United States," *Bulletin of the Atomic Scientists* 6 (Aug.–Sept. 1950): 266; Neal Fitzsimons, "Brief History of Civil Defense," in Eugene P. Wigner, ed., *Who Speaks for Civil Defense?* (New York: Charles Scribner's Sons, 1968), 31–33; Richard Polenberg, *War and Society: The United States, 1941–1945* (New York: J.B. Lippincott, 1972), 187; Richard R. Lingeman, *Don't You Know There's a War On?: The American Home Front, 1941–1945* (New York: G.P. Putnam's Sons, 1970), 34–37.

3. Merritt Abrash, "Through Logic to Apocalypse: Science-Fiction Scenarios of Nuclear Deterrence Breakdown," *Science-Fiction Studies* 13 (July 1986): 129.

4. Mauck, "History of Civil Defense," 268–69; the Shelter Program in Civil Defense, 1945–1969, attached to G. A. Lincoln to Glenn T. Seaborg, 26 May 1969, RG 326, U.S. Atomic Energy Commission, Secretariat Collection, Box 7764, MH & S 28 Civil Defense, U.S. Department of Energy Archives; Harry P. Yoshpe, "Our Missing Shield: The U.S. Civil Defense Program in Historical Perspective, Final Report for Federal Emergency Management Agency," Washington, D.C., Contract No. DCPA 01–79–C–0294, Apr. 1981, 84–87.

5. Mauck, "History of Civil Defense," 269; Yoshpe, "Our Missing Shield," 87–104; Statement by the President, 12 Jan. 1951, Box 1671, Official File, Papers of Harry S Truman.

6. J. Marshak, E. Teller, and L. R. Klein, "Dispersal of Cities and Industries," *Bulletin of the Atomic Scientists* 1 (15 Apr. 1946): 13; Robert A. Heinlein, "The Last Days of the United States," in *The New Worlds of Robert A. Heinlein: Expanded Universe* (New York: Grosset & Dunlap, 1980), 145–47, 156, 158; *The United States Strategic Bombing Survey: The Effects of Atomic Bombs on Hiroshima and Nagasaki* (Washington, D.C.: U.S. Government Printing Office, 1946), 41.

7. Tracy B. Augur, "The Dispersal of Cities as a Defense Measure," *Bulletin of the Atomic Scientists* 4 (May 1948): 131–34; Tracy B. Augur, "The Dispersal of Cities—A Feasible Program," *Bulletin of the Atomic Scientists* 4 (Oct. 1948): 312–15; Ralph E. Lapp, "The Strategy of Civil Defense," *Bulletin of the Atomic Scientists* 6 (Aug.–Sept. 1950): 241; "The Only Real Defense," *Bulletin of the Atomic Scientists* 7 (Sept. 1951): 242–43.

8. National Industrial Dispersion Policy, 10 Aug. 1951, Box 147, President's Secretary File, Papers of Harry S Truman; William L. C. Wheaton,

"Federal Action toward a National Dispersal Policy," *Bulletin of the Atomic Scientists* 7 (Sept. 1951): 271; Progress Report to the President, National Industrial Dispersion Program, 17 Jan. 1953, Box 147, President's Secretary File, Papers of Harry S Truman.

9. Lyon Gardiner Tyler, Jr., "Civil Defense: The Impact of the Planning Years, 1945–1950," Ph.D. dissertation, Duke University, 1967, 257–59.

10. Joseph G. Lemen to Martin Roess, 22 Sept. 1951, Box 20, Records of FCDA; Goulden, *The Best Years*, 266; clipping: "$938,000 Bomb Shelter Goes With Presidency," 11 Mar. 1952, Box 1671, Official File, Papers of Harry S Truman; Barney L. Allis to Matthew J. Connelly, 29 Jan. 1951, Box 1671, Official File, Papers of Harry S Truman; Tyler, "Civil Defense," 261, 282, 295–97, 303.

11. Statement by the President, 2 Nov. 1951, Box 227, President's Secretary File, Papers of Harry S Truman; Yoshpe, "Our Missing Shield," 164.

12. John A. DeChant to Governor Caldwell, 4 Apr. 1952, Box 19, Records of FCDA; Howard R. H. Johnson to Bess Landfear, 23 Apr. 1951, Box 32, Records of FCDA; John R. Steelman to Manly Fleischmann, 30 Oct. 1951, Box 1671, Official File, Papers of Harry S Truman; The Federal Civil Defense Audio-Visual Program, n.d., Box 5, Files of Spencer R. Quick, Papers of Harry S Truman; "Survival Under Atomic Attack," Box 1527, Official File, Papers of Harry S Truman.

13. The "Alert America" Convoys, 1 Nov. 1951, Box 1743, Official File, Papers of Harry S Truman; "THE CIVIL DEFENSE ALERT AMERICA CONVOY," n.d., Box 1743, Official File, Papers of Harry S Truman; "The Federal Civil Defense Administration presents SIGNS of our TIMES," 1952, Box 5, Files of Spencer R. Quick, Papers of Harry S Truman.

14. JoAnne Brown, "'A Is for *Atom*, B Is for *Bomb*': Civil Defense in American Public Education, 1948–1963," *Journal of American History* 75 (June 1988): 80–83.

15. *Bert the Turtle Says Duck and Cover*, Box 1, Files of Spencer R. Quick, Papers of Harry S. Truman; recording in personal possession; Brown, "'A Is for *Atom*,'" 83–84; Mary E. Meade, "What Programs of Civil Defense Are Needed in Our Schools?," *Bulletin of the National Association of Secondary-School Principals* 36 (Apr. 1952): 183; Grace Storm, "Civil Defense Film for Schools," *Elementary School Journal* 52 (Sept. 12, 1952): 12.

16. Press release of a letter from Harry S Truman to the speaker of the House of Representatives, 21 June 1951, Box 227, President's Secretary File, Papers of Harry S Truman; Address by Millard Caldwell, 15 May 1952, Box 5, Files of Spencer R. Quick, Papers of Harry S Truman.

17. "At Elm & Main," *Time*, 30 Mar. 1953, 58; "Greasewood Fires and Man's Most Terrible Weapon," 31; survey cited in Weart, *Nuclear Fear*, 132; Excerpts from President Eisenhower's Press Conference, Wednesday, 31 Mar. 1954, Energy History Collection, U.S. Department of Energy Archives; Divine, *Blowing on the Wind*, 12–13.

18. Report: Change in National Dispersion Policy, 5 November 1954, RG 326, U.S. Atomic Energy Commission, Thomas Murray Collection, Box 3710, Envelope B, U.S. Department of Energy Archives; Guy P. Jones to Pat Frank,

12 Apr. 1955, Box 86, Records of FCDA; Val Peterson quoted in Herbert Roback, "Civil Defense and National Defense," in Wigner, *Who Speaks for Civil Defense?*, 89; Mary M. Simpson, "A Long Hard Look at Civil Defense," *Bulletin of the Atomic Scientists* 12 (Nov. 1956): 346.

19. Dwight D. Eisenhower, *The White House Years: Mandate for Change, 1953–1956* (Garden City, N.Y.: Doubleday, 1963), 548–49; Stephen E. Ambrose, *Eisenhower*, Vol. 2, *The President* (New York: Simon & Schuster, 1984), 250–51.

20. James C. Hagerty, Press Release, 7 July 1955, Box 6, Whitman Name Series, Papers of Dwight D. Eisenhower; "Civil Defense: So Much to Be Done," *Newsweek*, 27 June 1955, 21–22; "Civil Defense: Best Defense? Prayer," *Time*, 27 June 1955, 17; "National Defense: Ducking for Cover," *Newsweek*, 30 July 1956, 28; Minutes of Cabinet Meeting, 25 July 1956, Box 7; Minutes of Cabinet Meeting, 12 July 1957, Box 9; Minutes of Cabinet Meeting, 19 July 1957, Box 9; Minutes of Cabinet Meeting, 9 Oct. 1958, Box 12—all above documents in Whitman Cabinet Series, Papers of Dwight D. Eisenhower.

21. Simpson, "A Long Hard Look at Civil Defense," 343–48; House Military Operations Subcommittee of the Committee on Government Operations, *Civil Defense for National Survival*, *Hearings before a Subcommittee of the Committee on Government Operations*, 84th Congress, 2nd session, 1956.

22. Dwight D. Eisenhower quoted in Herken, *Counsels of War*, 112–13, and in Rosenberg, "The Origins of Overkill," 40; Roback, "Civil Defense and National Defense," 89–90.

23. Herken, *Counsels of War*, 113–14; Clarfield and Wiecek, *Nuclear America*, 167–68; Robert Lovett quoted in Newhouse, *War and Peace in the Nuclear Age*, 119.

24. Dwight D. Eisenhower quoted in Newhouse, *War and Peace in the Nuclear Age*, 120; James R. Killian, Jr., *Sputnik, Scientists and Eisenhower: A Memoir of the First Special Assistant to the President for Science and Technology* (Cambridge, Mass.: MIT Press, 1977), 97; Dwight D. Eisenhower, *The White House Years: Waging Peace, 1956–1961* (Garden City, N.Y.: Doubleday, 1965), 221–23.

25. Address by Leo A. Hoegh, 2 Dec. 1958, Box 15, White House Office Staff Research Group Series, Papers of Dwight D. Eisenhower; Office of Civil and Defense Mobilization, Fact Sheet, Box 15, White House Office Staff Research Group Series, Papers of Dwight D. Eisenhower; the Shelter Program in Civil Defense, 1945–1969, attached to G. A. Lincoln to Glenn T. Seaborg, 26 May 1969, RG 326, U.S. Atomic Energy Commission, Secretariat Collection, Box 7764, MH & S 28 Civil Defense, U.S. Department of Energy Archives; Yoshpe, "Our Missing Shield," 164.

26. *Wall Street Journal*, 12 Feb. 1958; Leo A. Hoegh to General Goodpaster, 18 Sept. 1959, Box 15, White House Office Staff Research Group Series, Papers of Dwight D. Eisenhower; Office of Civil and Defense Mobilization, Information Bulletin, 10 Jan. 1961, RG 326, U.S. Atomic Energy Commission, Secretariat Collection, Box 1433, S & I 15—Civil Defense Folder, U.S. Department of Energy Archives; *How It Was Done*, Box 15, White House Office Staff Research Group Series, Papers of Dwight D. Eisenhower; Leo A. Hoegh to General Good-

paster, 13 Feb. 1959, Box 15, White House Office Staff Research Group Series, Papers of Dwight D. Eisenhower; Federal Civil Defense Administration, For Your Information, #510, 10 Apr. 1958, Box 9, White House Office Staff Research Group Series, Papers of Dwight D. Eisenhower.

27. "A Frightening Message for a Thanksgiving Issue," *Good Housekeeping*, Nov. 1958, 61; Editorial Comments on Shelter, Box 15, White House Office Staff Research Group Series, Papers of Dwight D. Eisenhower.

28. "West Coast Gets Ready," *Life*, 12 Mar. 1951, 64, 67–68; "H-Bomb Hideaway," *Life*, 23 May 1955, 169–70; Editorial Comments on Shelter, Box 15, White House Office Staff Research Group Series, Papers of Dwight D. Eisenhower; Office of Civil and Defense Mobilization, Information Bulletin, 10 Jan. 1961, RG 326, U.S. Atomic Energy Commission, Secretariat Collection, Box 1433, S & I 15—Civil Defense Folder, U.S. Department of Energy Archives.

29. "A Spare Room Fallout Shelter," *Life*, 25 Jan. 1960, 46; May, *Homeward Bound*, 103–7; Betty Friedan, *The Feminine Mystique* (New York: Dell, 1974), 236.

30. Ron Kovic, *Born on the Fourth of July* (New York: Pocket Books, 1977), 56; Leslie J. Miller, "Children's Hot Cold War," *New York Times*, 12 June 1982; Annie Dillard, *An American Childhood* (New York: Harper & Row, 1988), 181.

31. Schrag, "Seeing Ground Zero"; "Bomb Scares," *Newsweek*, 28 June 1982, 73; Tim O'Brien, *The Nuclear Age* (New York: Alfred A. Knopf, 1985), 9.

32. Guy P. Jones to Pat Frank, 12 Apr. 1955, Box 86, Records of FCDA; Pat Frank, *Alas, Babylon* (New York: Bantam, 1970), 137, 232.

33. Walter M. Miller, Jr., *A Canticle for Leibowitz* (New York: Bantam, 1980), 58, 245; Dowling, *Fictions of Nuclear Disaster*, 193–200; Brians, *Nuclear Holocausts*, 80–82, 260–61.

34. John F. Kennedy quoted in Henry Fairlie, *The Kennedy Promise* (New York: Dell, 1973), 57; John F. Kennedy quoted in Richard J. Walton, *Cold War and Counterrevolution: The Foreign Policy of John F. Kennedy* (Baltimore, Md.: Penguin, 1973), 9; John F. Kennedy, "Inaugural Address," 20 Jan. 1961, in *Public Papers of the Presidents of the United States: John F. Kennedy, 1961* (Washington, D.C.: U.S. Government Printing Office, 1962), 1.

35. Fred Kaplan, *The Wizards of Armageddon* (New York: Simon & Schuster, 1983), 307–9; John F. Kennedy, "Special Message to the Congress on Urgent National Needs," 25 May 1961, in *Public Papers of the Presidents: John F. Kennedy, 1961*, 402; Lapp, *Weapons Culture*, 42, 45; Steuart L. Pittman, "Government and Civil Defense," in Wigner, *Who Speaks for Civil Defense?*, 54.

36. Allan M. Winkler, *Modern America: The United States from World War II to the Present* (New York: Harper & Row, 1985), 117; Kaplan, *The Wizards of Armageddon*, 291–93; Herbert S. Parmet, *JFK: The Presidency of John F. Kennedy* (New York: Dial Press, 1983), 181, 189–90.

37. Parmet, *JFK*, 193, 196, 197; Michael Mandelbaum, *The Nuclear Question: The United States & Nuclear Weapons, 1946–1976* (New York: Cambridge Univ. Press, 1979), 94; John F. Kennedy, "Radio and Television Report to the American People on the Berlin Crisis," 25 July 1961, in *Public Papers of the*

Presidents: John F. Kennedy, 1961, 534–35; "All Out Against Fallout," *Time*, 4 Aug. 1961, 11; "Survival: Are Shelters the Answer?," *Newsweek* 6 Nov. 1961, p. 21.

38. John F. Kennedy, "Radio and Television Report to the American People on the Berlin Crisis," p. 536; "All Out Against Fallout," 11.

39. "A Message to You from the President," and "Fallout Shelters," *Life*, 15 Sept. 1961, 95; Pittman, "Government and Civil Defense," 55, 61–62, 68–70; John F. Kennedy quoted in Parmet, *JFK*, 198; Exercise Spade Fork, Sept. 6–27, 1962: Civil Evaluation Report, 30 Jan. 1963, NLK–78–606, in Papers of John F. Kennedy, John F. Kennedy Library, Boston, Mass.; Exercise Spade Fork: Evaluation Report, 1 Feb. 1963, NLK–78–607, Papers of John F. Kennedy; *Fallout Protection: What to Know and Do about Nuclear Attack*, Box 75, Papers of Sidney R. Yates.

40. "Fallout Shelters: Low-Key Plan," *Newsweek*, 8 Jan. 1962, 18; "Shelters; Jitters Ease, Debate Goes On," *Newsweek*, 15 Jan. 1962, 50–51; Pittman, "Government and Civil Defense," 66–67; Carl Kaysen to Mr. Bundy, 3 Nov. 1961, NLK–78–641, Box 295, National Security Files: Civil Defense, 10/28/61–11/12/61, Papers of John F. Kennedy.

41. "Civil Defense: The Sheltered Life," *Time*, 20 Oct. 1961, 21; "Let's Prepare . . . Shelters," *Life*, 13 Oct. 1961, 4; "All Out Against Fallout," 11; "Civil Defense: Survival (Contd.)," *Time*, 3 Nov. 1961, 19; Richard Russell quoted in *New York Times*, 22 Jan. 1971.

42. "Building: Shelter Skelter," *Time*, 1 Sept. 1961, 59–60; "Fallout Shelters," 98–108.

43. Lawrence S. Wittner, *Cold War America: From Hiroshima to Watergate*, Expanded Edition (New York: Holt, Rinehart and Winston, 1978), 215; J. Samuel Walker, Commentary in session on Problems of the Atomic Age, Organization of American Historians Convention, Cincinnati, Ohio, 9 Apr. 1983; Silber, *Test and Protest*, 119.

44. "Civil Defense: A Place to Hide," *Time*, 18 Dec. 1950, 21; "Religion: Gun Thy Neighbor?," *Time*, 18 Aug. 1961, 58; "Love Thy Neighbor," *Newsweek*, 9 Oct. 1961, 94; "The Debate Goes On," *Newsweek*, 4 Dec. 1961, 16; Lapp, *The Weapons Culture*, 50.

45. *The Effects of Nuclear War*, a report from the National Committee for a Sane Nuclear Policy, n.d., Box 10, Folder: Literature, 1957–1962; *How Sane Are Fallout Shelters?*, Box 20—both documents in Records of the National Committee for a Sane Nuclear Policy.

46. "Civil Defense: Coffins or Shields?," *Time*, 2 Feb. 1962, 15; Wittner, *Rebels against War*, 266; *New York Times*, 2 Sept. 1961.

47. "Civil Defense: The Sheltered Life," 21.

48. Federation of American Scientists, Press Release, 4 Dec. 1961, Box 75, Papers of Sidney R. Yates; Gerard Piel, *The Illusion of Civil Defense*, 10 Nov. 1961, p. 21, Box 75, Papers of Sidney R. Yates.

49. Treaty Banning Nuclear Weapons Tests in the Atmosphere, in Outer Space and Under Water, in Glenn Seaborg, *Kennedy, Krushchev, and the Test Ban* (Berkeley and Los Angeles: Univ. of California Press, 1983), Appendix, 302–5.

50. Lyndon B. Johnson and Gerald Ford quoted in Fred M. Kaplan, "The

Soviet Civil Defense Myth, Part 2," *Bulletin of the Atomic Scientists* 34 (Apr. 1978): 45.

51. Roback, "Civil Defense and National Defense," 99–100; "Civil Defense: Are Federal, State, and Local Governments Prepared for Nuclear Attack?," Report LCD–76–464 by the comptroller general of the United States, 8 Aug. 1977, 32, 34.

52. Yoshpe, "Our Missing Shield," 406, 467, 480–81, 486–90; "Civil Defense: Are Federal, State, and Local Governments Prepared for Nuclear Attack?," 1.

53. Robert Scheer, *With Enough Shovels: Reagan, Bush, and Nuclear War* (New York: Random House, 1982), 18, 21; Thomas J. Kerr, *Civil Defense in the U.S.: Bandaid for a Holocaust?* (Boulder, Colo.: Westview Press, 1983), 165–67; "Fallout Shelters: Making a Comeback," *Newsweek*, 22 Feb. 1982, 10.

54. Jennifer Leaning, *Civil Defense in the Nuclear Age: What Purpose Does It Serve and What Survival Does It Promise?* (Cambridge, Mass.: Physicians for Social Responsibility, 1982), 8; Richard M. Ketchum, "An Uncivil Defense," *The Reader's Digest*, Feb. 1983, 99; Louis O. Guiffrida quoted in Bernard Weintraub, "Civil Defense Agency Suddenly in Spotlight," *International Herald-Tribune*, 13 Apr. 1982; "Fallout Shelters: Making a Comeback," 10; "Does Civil Defense Make Sense?," *Newsweek*, 26 Apr. 1982, 33.

55. Ellen Goodman, "Civil Defense Is Really Military Madness," *Eugene* [Oregon] *Register-Guard*, 2 Feb. 1982; Ellen Goodman, "A Big Hand, Folks, for the Nuclear Follies," *Eugene* [Oregon] *Register-Guard*, 10 Dec. 1983; Fred Small, *Breaking from the Line: The Songs of Fred Small* (Cambridge, Mass.: Yellow Moon Press, 1986), 62–63; Edward Markey quoted in Tom Wicker, "Civil Defense a Dangerous Band-Aid," *Eugene* [Oregon] *Register-Guard*, 2 Aug. 1982.

56. George Kennan quoted in Arthur M. Schlesinger, Jr., Reflections on Civil Defense, n.d., NLK–77–701, Box 295, NSF File, Subjects: Civil Defense, 12/61, Papers of John F. Kennedy.

6. The Peaceful Atom

1. [Harry S Truman], The Significance of the Atomic Age, 15 Oct. 1945, Box 112, President's Secretary File, Papers of Harry S Truman; David Lilienthal and Robert M. Hutchins quoted in Daniel Ford, *The Cult of the Atom: The Secret Papers of the Atomic Energy Commission* (New York: Simon & Schuster, 1982), 30–31; The phrase "too cheap to meter" at the end of the paragraph was used by Lewis Strauss in a speech in Sept. 1954 and is also quoted in Ford, *The Cult of the Atom*, 50; George Gamow, *Atomic Energy in Cosmic and Human Life* (New York: Macmillan, 1946), ix; David Dietz, *Atomic Energy in the Coming Era* (New York: Dodd, Mead & Co., 1945), 13, 16–17; "Pulling the Nuclear Plug," *Time*, 13 Feb. 1984, 34.

2. Gordon Dean, The Atom in National Defense—Remarks to the American Bar Association, San Francisco, Calif., 17 Sept. 1952, Box 1530, Official File, Papers of Harry S Truman; Gordon Dean, "Today: Atoms *Work* for You,"

Parade, 24 Feb. 1952, 6–8; Lewis Strauss, "My Faith in the Atomic Future," *The Reader's Digest*, Aug. 1955, 17; John A. McCone, The Atom: Protector, Destroyer, Benefactor—Remarks at Rockhurst College, Kansas City, Mo., 16 Apr. 1959, Document 12298, Records of the Department of Energy, Coordination and Information Center; Glenn Seaborg quoted in Ford, *The Cult of the Atom*, 23–24.

3. Boyer, *By the Bomb's Early Light*, 296–97; Kurt Vonnegut, Jr., *Player Piano* (New York: Bard, 1969), 58.

4. Ira M. Freeman, *All About the Atom* (New York: Random House, 1955), jacket; Carlton Pearl, *The Tenth Wonder: Atomic Energy* (Boston: Little, Brown, 1956), xi; for other children's accounts, see Mae and Ira Freeman, *The Story of the Atom* (New York: Random House, 1960); and Nelson F. Bealer and Franklyn M. Branley, *Experiments with Atomics* (New York: Thomas Y. Crowell, 1965).

5. Haber, *The Walt Disney Story of Our Friend the Atom*, 13–21.

6. *Operation Atomic Vision* quoted in Boyer, *By the Bomb's Early Light*, 298–99; Brown, "'A Is for *Atom*,'" 72–73.

7. Erskine, "The Polls," 179; Charles A. Metzner and Julia B. Kessler, "What Are the People Thinking?," *Bulletin of the Atomic Scientists* 7 (Nov. 1951): 341, 352; Central Surveys—A P.I.P. Report to Member Companies, File: Atomic Power, 12 Aug. 1953, Box 523, Official File, Central File, Papers of Dwight D. Eisenhower.

8. May, *Homeward Bound*, 110–12; "Atomic Cafe."

9. Pearl, *The Tenth Wonder*, 84–85; Hilgartner, Bell, and O'Connor, *Nukespeak*, 49–50; ABOUT PLOWSHARE . . . a Backgrounder, 6 Mar. 1970, Box 21, Records of the Committee for a Sane Nuclear Policy; *Plowshare*, U.S. Atomic Energy Commission, 1966, Document 15316, Records of the Department of Energy, Coordination and Information Center; Peter Coates, "Project Chariot: Alaskan Roots of Environmentalism," *Alaska History* (Fall 1989), 1–2; Walker, "Writing the History of Nuclear Energy," 378–79.

10. Mazuzan and Walker, *Controlling the Atom*, 3–4; Mazuzan and Trask, "An Outline History," 8; Steven L. Del Sesto, *Science, Politics, and Controversy: Civilian Nuclear Power in the United States, 1946–1974* (Boulder, Colo.: Westview, 1979), 17–24; Michael J. Brenner, *Nuclear Power and Non-Proliferation: The Remaking of U.S. Policy* (New York: Cambridge Univ. Press, 1981), 3–4.

11. Ford, *The Cult of the Atom*, 32–33; George T. Mazuzan, "Conflict of Interest: Promoting and Regulating the Infant Nuclear Power Industry, 1954–1956," *The Historian* 44 (Nov. 1981): 2; Balogh, *Chain Reaction*, 95.

12. David Lilienthal quoted in Del Sesto, *Science, Politics, and Controversy*, 40; J. Robert Oppenheimer quoted in Mazuzan and Trask, "An Outline History," 9; Ford, *The Cult of the Atom*, 33–34.

13. *EBR–1* and *Self Guided Tour of ERB–1*, pamphlets distributed at the reactor site; Jack M. Holl, Roger M. Anders, and Alice L. Buck, *United States Civilian Nuclear Power Policy, 1954–1984: A Summary History* (Washington, D.C.: U.S. Department of Energy, Feb. 1986), 2.

14. Richard G. Hewlett and Jack M. Holl, *Atoms for Peace and War, 1953–1962: Eisenhower and the Atomic Energy Commission* (Berkeley and Los Angeles:

Univ. of California Press, 1989), 15, 21; Holl, Anders, and Buck, *Civilian Nuclear Power Policy*, 1.

15. Hewlett and Holl, *Atoms for Peace and War*, 42–44, 62, 65–66, 71–72; CHRONOLOGY—"Atoms for Peace" Project, 30 Sept. 1954, Box 24, C. D. Jackson Papers, Dwight D. Eisenhower Library.

16. Dwight D. Eisenhower, "Address Before the General Assembly of the United Nations on Peaceful Uses of Atomic Energy, New York City," 8 Dec. 1953, in *Public Papers of the Presidents of the United States: Dwight D. Eisenhower, 1953* (Washington, D.C.: U.S. Government Printing Office, 1960), 813–22; Hewlett and Holl, *Atoms for Peace and War*, 72, 209.

17. Holl, Anders, and Buck, *Civilian Nuclear Power Policy*, 2; Clarfield and Wiecek, *Nuclear America*, 185; Atomic Power Abroad, National Security Council, NSC 5507, 28 Jan. 1955; Progress Report on Nuclear Energy Projects and Related Information Programs (including NSC 5507/2) by the Operations Coordinating Board, National Security Council, 21 Dec. 1955; Progress Report on Peaceful Uses of Atomic Energy (NSC 5507/2) by the Department of State and the Atomic Energy Commission, National Security Council, 22 Apr. 1957—all documents in the Dwight D. Eisenhower Library.

18. Holl, Anders, and Buck, *Civilian Nuclear Power Policy*, 1–2; Mazuzan, "Conflict of Interest," 2–3; Lewis Strauss quoted in Ford, *The Cult of the Atom*, 41.

19. Holl, Anders, and Buck, *Civilian Nuclear Power Policy*, 3; Ford, *The Cult of the Atom*, 41–42; Willard Libby quoted in Mazuzan, "Conflict of Interest," 4.   20. Richard G. Hewlett and Francis Duncan, *Nuclear Navy, 1946–1962* (Chicago: The Univ. of Chicago Press, 1974), 32–35; Richard G. Hewlett and Francis Duncan, *Atomic Shield*, 75.

21. Hewlett and Holl, *Atoms for Peace and War*, 186–87.

22. Ford, *The Cult of the Atom*, 53, 58, 61, 66; Hewlett and Duncan, *Nuclear Navy*, 97–100, 109–17; Hewlett and Holl, *Atoms for Peace and War*, 187–88.

23. Hewlett and Holl, *Atoms for Peace and War*, 192, 196–97, 227–28; Mazuzan and Walker, *Controlling the Atom*, 21.

24. Hewlett and Holl, *Atoms for Peace and War*, 419–22.

25. Ibid., 201–2, 205–8, 420–21; Holl, Anders, and Buck, *Civilian Nuclear Power Policy*, 4–5; Report to the General Manager by a Special Task Force, Atomic Energy Commission, Policy Relating to Peaceful Power Reactor Uses of Atomic Energy, AEC 655/40, 28 Nov. 1955, RG 326, U.S. Atomic Energy Commission, Secretariat Collection, Box 1245, IRA6 General Policy Folder, U.S. Department of Energy Archives.

26. Holl, Anders, and Buck, *Civilian Nuclear Power Policy*, 5; Hewlett and Holl, *Atoms for Peace and War*, 341–45.

27. Joint Committee on Atomic Energy, *Proposed Expanded Civilian Nuclear Power Program* (Washington, D.C.: U.S. Government Printing Office, 1958), 1–2; 8–9; Holl, Anders, and Buck, *Civilian Nuclear Power Policy*, 8.

28. Ford, *The Cult of the Atom*, 58; "Atomic Power: Cinderella Is Slipping Back into the Kitchen," *Science* 136 (20 Apr. 1962): 244.

29. U.S. Atomic Energy Commission, *Civilian Nuclear Power . . . A Report to the President—1962*, 8, 10, 26, 34, 43.

30. Holl, Anders, and Buck, *Civilian Nuclear Power Policy*, 11.

31. The discussion of nuclear power and its problems in the rest of this chapter draws on Winkler, "The Nuclear Question," 141–45. See also Del Sesto, *Science, Politics, and Controversy*, 85–86; Ford, *The Cult of the Atom*, 62–63.

32. Walker, *Containing the Atom*, 391–92; "The Atlantic Report: Science and Industry," *Atlantic Monthly* 213 (Mar. 1964): 22, 25–26, 30; Carl Dreher, "Atomic Power: The Fear and the Promise," *The Nation* 198 (23 Mar. 1964): 289–93; Joseph P. Blank, "Atomic Power Comes of Age," *The Reader's Digest* 87 (Dec. 1965): 112; "Are Nuclear Plants Winning Acceptance?," *Electrical World* 165 (24 Jan. 1966): 115–18; "Nuclear Power and the Community: Familiarity Breeds Confidence," *Nuclear News* 9 (May 1966): 15–16.

33. Del Sesto, *Science, Politics, and Controversy*, 91; Fritz F. Heimann, "How Can We Get the Nuclear Job Done?," in Arthur W. Murphy, ed., *The Nuclear Power Controversy* (Englewood Cliffs, N.J.: Prentice Hall, 1976), 95–96; Holl, Anders, and Buck, *Civilian Nuclear Power Policy*, 22.

34. Mazuzan, "Conflict of Interest," 5–6; Edward Teller with Allen Brown, *The Legacy of Hiroshima* (Garden City, N.Y.: Doubleday, 1962), 102.

35. Mazuzan and Walker, *Controlling the Atom*, 59–65; Lewis Strauss quoted in Mazuzan, "Conflict of Interest," 8.

36. *Governmental Indemnity: Hearings before the Joint Committee on Atomic Energy on Governmental Indemnity for Private Licensees and AEC Contractors against Reactor Hazards*, 84th Cong., 2nd sess., 1956, 46, 49; Mazuzan, "Conflict of Interest," 10–11.

37. U.S. Atomic Energy Commission, *Theoretical Possibilities and Consequences of Major Accidents in Large Nuclear Power Plants*, WASH–740, Mar. 1957 (Washington, D.C.: U.S. Atomic Energy Commission, 1957), 1, 13–14; Mazuzan and Walker, *Controlling the Atom*, 203–13.

38. Mazuzan and Walker, *Controlling the Atom*, 16, 80, 124–34; Hewlett and Holl, *Atoms for Peace and War*, 352, 358; George T. Mazuzan and J. Samuel Walker, "The Safety Goal Issue in Historical Perspective," Dec. 1980, 1–2, in Historical Office, Office of the Secretary, U.S. Nuclear Regulatory Commission.

39. Chauncy Starr quoted in Mazuzan and Walker, "The Safety Goal Issue in Historical Perspective," 10.

40. Dwight A. Ink to Christopher H. Russell, 10 Jan. 1961, Box 2, White House Office, Staff Research Group, Papers of Dwight D. Eisenhower; Walter C. Patterson, *Nuclear Power* (New York: Penguin, 1976), 175–76, 180–82; Sheldon Novick, *The Careless Atom* (Boston: Houghton Mifflin, 1969), 159–60; McKinley C. Olson, *Unacceptable Risk: The Nuclear Power Controversy* (New York: Bantam, 1976), 21, 62–64; John G. Fuller, "We Almost Lost Detroit," in Peter Faulkner, ed., *The Silent Bomb: A Guide to the Nuclear Energy Controversy* (New York: Vintage, 1977), 45–59; Walker, *Containing the Atom*, 167–68.

41. Ford, *The Cult of the Atom*, 66–67.

42. Walker, *Containing the Atom*, 203–212, 214–219.

43. *Reactor Safety Study: An Assessment of Risks in U.S. Commercial Nuclear Power Plants*, WASH–1400 (NUREG–75/014), Oct. 1975 (Springfield, Va.: U.S. Department of Commerce, National Technical Information Service, 1975); David Bodansky and Fred H. Schmidt, "Safety Aspects of Nuclear Energy," in Murphy, *The Nuclear Power Controversy*, 37; Ford, *The Cult of the Atom*, 138–43.

44. Executive Summary (Appendix 1), in Vol. 1, *Reactor Safety Study*, 9;

Bodansky and Schmidt, "Safety Aspects," 37–39; California Assembly, Committee on Resources, Land Use and Energy, "Reactor Safety," in Faulkner, *The Silent Bomb*, 145–50; Olson, *Unacceptable Risk*, 22–24; Ford, *The Cult of the Atom*, 143–53.

45. Ford, *The Cult of the Atom*, 154–69; "Rasmussen Issues Revised Odds on a Nuclear Catastrophe," *Science* 190 (14 Nov. 1975): 640; "Nuclear Reactor Safety—the APS Submits Its Report," *Physics Today* 28 (July 1975): 39, 41–42; Norman C. Rasmussen, "The Safety Study and its Feedback," *Bulletin of the Atomic Scientists* 31 (Sept. 1975): 25–28.

46. Hilgartner, Bell, and O'Connor, *Nukespeak*, 140.

47. *Report of the President's Commission on the Accident at Three Mile Island— The Need for Change: The Legacy of TMI* (Washington, D.C.: U.S. Government Printing Office, 1979), 101–61; Charles Perrow, *Normal Accidents: Living with High-Risk Technologies* (New York: Basic Books, 1984), 15–31; Clarfield and Wiecek, *Nuclear America*, 384–85; "A Nuclear Nightmare," *Time*, 9 Apr. 1979, 8, 11–12, 15–16, 19; "Nuclear Accident," *Newsweek*, 9 Apr. 1979, 24–30, 33.

48. Cynthia Bullock Flynn, "Reactions of Local Residents to the Accident at Three Mile Island," in David L. Sills, C. P. Wolf, and Vivien B. Shelanski, eds., *Accident at Three Mile Island: The Human Dimensions* (Boulder, Colo.: Westview, 1982), 49–56; Lonna Malmsheimer, "Three Mile Island: Fact, Frame and Fiction," *American Quarterly* 38 (Spring 1986): 37–38, 42–49; Anne Tyler, *Dinner at the Homesick Restaurant* (New York: Berkeley Books, 1983), 272.

49. *Report of the President's Commission on the Accident at Three Mile Island*, 2, 10, 27; Ford, *The Cult of the Atom* 230–32.

50. "Meltdown," *Newsweek* 12 May 1986, 20, 22–23; Steve Emmons, "Sick Jokes: Coping With the Horror," *Los Angeles Times*, 30 May 1986; "More Fallout From Chernobyl," *Time*, 19 May 1986, 44–45; Anthony Lewis, "Lessons of Chernobyl Lost on Leaders," *Eugene* [Oregon] *Register-Guard*, 30 May 1986; "Chernobyl's Cancer Toll Will Surpass Million, a Berkeley Physicist Predicts," *Chronicle of Higher Education*, 17 Sept. 1986, 5; William J. Broad, "A Mountain of Trouble," *New York Times Magazine*, 18 Nov. 1990, 39; Joyce Maynard, "The Story of a Town," *New York Times Magazine*, 11 May 1986, 42.

51. J. Samuel Walker, "Nuclear Power and the Environment: The Atomic Energy Commission and Thermal Pollution, 1965–1971," *Technology and Culture* 30 (Oct. 1989): 965–66, 968–73, 975–79, 983–85, 987–89; Balogh, *Chain Reaction*, 258–65; Robert H. Boyle, "The Nukes Are in Hot Water," *Sports Illustrated*, 20 Jan. 1969, 24–28.

52. Del Sesto, *Science, Politics, and Controversy*, 157–59; Clarfield and Wiecek, *Nuclear America*, 358–59; *Calvert Cliffs Coordinating Committee* v. *AEC*, 449 F.2d 1109, 1971; Holl, Anders, and Buck, *Civilian Nuclear Power Policy*, 15.

53. Barbara Epstein, *Political Protest and Cultural Revolution: Nonviolent Direct Action in the 1970s and 1980s* (Berkeley and Los Angeles: Univ. of California Press, 1991), 9–10.

54. Ibid., 11–12, 105.

55. Holl, Anders, and Buck, *Civilian Nuclear Power Policy*, 19–20; Maynard, "The Story of a Town," 20, 22, 24, 26, 40–43, 47, 50; Broad, "A Mountain of Trouble," 37.

56. Holl, Anders, and Buck, *Civilian Nuclear Power Policy*, 17, 22; Balogh, *Chain Reaction*, 237.

57. "Whoops! A $2 Billion Blunder," *Time*, 8 Aug. 1983, 50–52; *New York Times*, 14 Aug. 1983, 22 Jan. 1984; "Pulling the Nuclear Plug," 34–35; "A Meltdown for Nuclear Power," *Business Week*, 30 Jan. 1984, 18–19; *Eugene* [Oregon] *Register-Guard*, 30 Aug. 1983, 14 Jan. 1984; *Cincinnati Enquirer*, 15 Mar. 1987; Harold P. Green, "The Peculiar Politics of Nuclear Power," *Bulletin of the Atomic Scientists* 38 (Dec. 1982): 63–64; James Cook, "Nuclear Follies," *Forbes*, 11 Feb. 1985, cover; Michael Mariotte, "Sudden Burst of Energy," *Nuclear Times*, July–Aug. 1986, 16; Ronald Reagan, "Nuclear Energy Policy," 8 Oct. 1981, *Weekly Compilation of Presidential Documents* 17 (12 Oct. 1981), 1101.

58. On the merging of images, see Weart, *Nuclear Images*, 320–21; Alvin M. Weinberg, "Social Institutions and Nuclear Energy," *Science* 177 (7 July 1972): 33.

7. The Search for Stability

1. Dwight D. Eisenhower, quoted in John Lewis Gaddis, *Strategies of Containment: A Critical Appraisal of Postwar American National Security Policy* (New York: Oxford Univ. Press, 1982), 150; Bulletin of the Atomic Scientists, "We Tell the World What Time It Is" (poster).

2. Freeman Dyson, *Disturbing the Universe* (New York: Harper & Row, 1979), 144–45; "The Talk of the Town," *The New Yorker*, 21 Jan. 1985, 21–22.

3. Herken, *Counsels of War*, 6, 16–17, 47, 49.

4. Paul H. Nitze, "Atoms, Strategy and Policy," *Foreign Affairs* 34 (Jan. 1956): 187–88, 195–96, 198.

5. Herken, *Counsels of War*, 88–89, 91, 111.

6. Albert Wohlstetter, "The Delicate Balance of Terror," *Foreign Affairs* 37 (Jan. 1959): 211–12, 215, 220–21, 230.

7. "General Taylor—Questions U.S. Army Policy," *U.S. News & World Report*, 6 July 1959, 23; Maxwell D. Taylor, *The Uncertain Trumpet* (New York: Harper & Brothers, 1960), 5–8, 17, 23, 30–34, 139, 158–61.

8. John F. Kennedy quoted in Newhouse, *War and Peace in the Nuclear Age*, 162; David Halberstam, *The Best and the Brightest* (New York: Random House, 1972).

9. Halberstam, *The Best and the Brightest*, 51–63; Bundy, *Danger and Survival*, 352, 354; McGeorge Bundy quoted in Kaplan, *The Wizards of Armageddon*, 297; Scott D. Sagan, "SIOP–62: The Nuclear War Plan Briefing to President Kennedy," *International Security* 12 (Summer 1987): 23.

10. Halberstam, *The Best and the Brightest*, 220–21, 225–32; Herken, *Counsels of War*, 146.

11. Robert McNamara quoted in Theodore C. Sorensen, *Kennedy* (New York: Harper & Row, 1965), 603; Herken, *Counsels of War*, 146–47, 156.

12. Michael Mandelbaum, *The Nuclear Revolution: International Politics be-*

fore and after Hiroshima (New York: Cambridge Univ. Press, 1981), 47; Robert McNamara quoted in Michael Mandelbaum, *The Nuclear Question: The United States and Nuclear Weapons, 1946–1976* (New York: Cambridge Univ. Press, 1979), 90; Sorensen, *Kennedy*, 626–27; Gaddis, *Strategies of Containment*, 199.

13. Kaplan, *The Wizards of Armageddon*, 299–300; Newhouse, *War and Peace in the Nuclear Age*, 163–65; Robert McNamara quoted in William W. Kaufmann, *The McNamara Strategy*, (New York: Harper & Row, 1964), 116.

14. Herken, *Counsels of War*, 173; Sorensen, *Kennedy*, 627; Mandelbaum, *The Nuclear Question*, 124–25; Newhouse, *War and Peace in the Nuclear Age*, 164; Friedberg, "A History of the U.S. Strategic 'Doctrine,' " 50–51; the Harvard Nuclear Study Group, *Living with Nuclear Weapons*, 88–89; Winkler, "The Nuclear Question," 136–37.

15. Kaplan, *The Wizards of Armageddon*, 302–4; Mandelbaum, *The Nuclear Question*, 95–96.

16. Winkler, *Modern America*, 118; Stewart Alsop and Charles Bartlett, "In Time of Crisis," *The Saturday Evening Post*, 8 Dec. 1962, 16; "Introduction by Harold Macmillan" in Robert F. Kennedy, *Thirteen Days: A Memoir of the Cuban Missile Crisis* (New York: W.W. Norton, 1969), 17.

17. Sorensen, *Kennedy*, 682; Kaplan, *The Wizards of Armageddon*, 304–5; Winkler, *Modern America*, 118.

18. Entire conversation quoted in Kaplan, *The Wizards of Armageddon*, 305.

19. Sorensen, *Kennedy*, 703–4, 707–18, John F. Kennedy, "Radio and Television Report to the American People on the Soviet Arms Buildup in Cuba," 22 Oct. 1962, in *Public Papers of the Presidents of the United States: John F. Kennedy, 1962* (Washington, D.C.: U.S. Government Printing Office, 1963), 807–8.

20. Winkler, *Modern America*, 119; Dean Acheson, "Dean Acheson's Version of Robert Kennedy's Version of the Cuban Missile Affair: Homage to Plain Dumb Luck," *Esquire*, Feb. 1969, 76.

21. Kaplan, *The Wizards of Armageddon*, 305–6.

22. "Fall 1961," in Robert Lowell, *Selected Poems* (New York: Farrar, Straus & Giroux, 1976), 105; Lehrer, *Too Many Songs by Tom Lehrer*, 156–57; Tom Lehrer, "That Was the Year that Was," R–6179, Reprise Records, n.d.

23. Eugene Burdick and Harvey Wheeler, *Fail-Safe* (New York: Dell, 1962), preface; Michael G. Wollscheidt, "Fail Safe," in Shaheen, *Nuclear War Films*, 68–71; Winkler, "American Hopes and Fears in the Nuclear Age," 34–35.

24. Burdick and Wheeler, *Fail-Safe*, 278.

25. Wollscheidt, "Fail-Safe," 68, 74–75.

26. George W. Linden, "Dr. Strangelove or: How I Learned to Stop Worrying and Love the Bomb," in Shaheen, *Nuclear War Films*, 60, 62–65.

27. Ibid., 58, 60; Robert B. Tucker, "From Doom-and-Gloom to 'The Coming Boom,' " *United: The Magazine of the Friendly Skies*, July 1983, 92; Peter George, *Dr. Strangelove or: How I Learned to Stop Worrying and Love the Bomb* (New York: Bantam, 1963), 98.

28. Bundy, *Danger and Survival*, 460.

29. John F. Kennedy quoted in Lapp, *Weapons Culture*, 41, and in Anthony Lewis, "Kennedy Showed That He Could Learn," *Eugene* [Oregon] *Register-*

Guard, 16 Nov. 1983; Chester Bowles, *Promises to Keep: My Years in Public Life, 1941–1969* (New York: Harper & Row, 1971), 452.

30. Nikita Khrushchev and John Kennedy quoted in Seaborg, *Kennedy, Khrushchev, and the Test Ban*, 176, and in Parmet, *JFK*, 310–11; Nikita Khrushchev, "Report to the Supreme Soviet of the USSR," 12 Dec. 1962, from *Pravda*, 13 Dec. 1962, in Alexander Dallin, ed., *Diversity in International Communism: A Documentary Record, 1961–1963* (New York: Columbia Univ. Press, 1963), 676.

31. Mandelbaum, *The Nuclear Question*, 163–65, 168.

32. Ibid., 172; Parmet, *JFK*, 312; Seaborg, *Kennedy, Khrushchev, and the Test Ban*, 212.

33. John F. Kennedy, "Commencement Address at American University in Washington," 10 June 1963, in *Public Papers of the Presidents: John F. Kennedy, 1963* (Washington, D.C.: U.S. Government Printing Office, 1964), 460–64.

34. The Manchester *Guardian* and Nikita Khrushchev quoted in Sorensen, *Kennedy*, 733; Seaborg, *Kennedy, Khrushchev, and the Test Ban*, 217–18.

35. Parmet, *JFK*, 313; Halberstam, *The Best and the Brightest*, 73–74, 90, 192–97; Mandelbaum, *The Nuclear Question*, 175–76; Arthur M. Schlesinger, Jr., *A Thousand Days: John F. Kennedy in the White House* (Boston: Houghton Mifflin, 1965), 903.

36. Theodore Sorensen quoted in Mandelbaum, *The Nuclear Question*, 175.

37. Parmet, *JFK*, 314; Mandelbaum, *The Nuclear Question*, 177, 179; John F. Kennedy quoted in Stephen Ambrose, *Rise to Globalism: American Foreign Policy, 1938–1970* (Baltimore: Penguin, 1971), 294.

38. Edward Teller quoted in Parmet, *JFK*, 315; Herbert York, *Race to Oblivion* (New York: Simon & Schuster, 1970), 44–45; Bulletin of the Atomic Scientists, "We Tell the World What Time It Is" (poster).

39. Herman Kahn quoted in Hodgson, *America in Our Time*, 8; Mandelbaum, *The Nuclear Question*, 192–94; Stanford Arms Control Group, *International Arms Control: Issues and Agreements*, edited by Coit D. Blacker and Gloria Duffy (Stanford, Calif.: Stanford Univ. Press, 1984), 154–58; Bulletin of the Atomic Scientists, "We Tell the World What Time It Is" (poster).

40. Winkler, *Modern America*, 166–67, 169–70; Winkler, "The Nuclear Question," 147; Mandelbaum, *The Nuclear Revolution*, 194; Mandelbaum, *The Nuclear Question*, 199; Stanford Arms Control Group, *International Arms Control*, 225–26, 245–46.

41. Newhouse, *War and Peace in the Nuclear Age*, 220, 230–31.

42. Bundy, *Danger and Survival*, 549–50; Winkler, "The Nuclear Question," 135.

43. Newhouse, *War and Peace in the Nuclear Age*, 221; Mandelbaum, *The Nuclear Question*, 193; "SALT I Accords, 1972," in Stanford Arms Control Group, *International Arms Control*, 413–33; Stanford Arms Control Group, *International Arms Control*, 229–45, 249–54; Harvard Nuclear Study Group, *Living with Nuclear Weapons*, 92–95.

44. Henry Kissinger, *White House Years* (Boston: Little, Brown, 1979), 1245; Gaddis, *Strategies of Containment*, 325; Henry Kissinger quoted in Gerard Smith, *Doubletalk: The Story of the First Strategic Arms Limitation Talks* (Garden City, N.Y.:

Doubleday, 1980), 177; Nigel Calder, *Nuclear Nightmares: An Investigation into Possible Wars* (New York: Penguin, 1981), 58.

45. Jimmy Carter, "Address before the General Assembly," 4 Oct. 1977, in *Public Papers of the Presidents of the United States: Jimmy Carter, 1971*, Book 2 (Washington, D.C.: U.S. Government Printing Office, 1978), 1716; "SALT II Treaty, 1979," in Stanford Arms Control Group, *International Arms Control*, 446–77; Stanford Arms Control Group, *International Arms Control*, 260–67; Harvard Nuclear Study Group, *Living with Nuclear Weapons*, 99; Calder, *Nuclear Nightmares*, 120.

46. Stanford Arms Control Group, *International Arms Control*, 267–72; Newhouse, *War and Peace in the Nuclear Age*, 295–96; "SALT II: Sealed with a Kiss," *Newsweek*, 2 July 1979, 25–26; Harvard Nuclear Study Group, *Living with Nuclear Weapons*, 98–99.

47. Stanford Arms Control Group, *International Arms Control*, 272–76; Harvard Nuclear Study Group, *Living with Nuclear Weapons*, 196, 206, 211; Newhouse, *War and Peace in the Nuclear Age*, 340–45.

8. A Resurgence of Concern

1. H. Jack Geiger quoted in "Thinking the Unthinkable," *Newsweek*, 5 Oct. 1981, 38; Pentagon analyst quoted in Carol Cohn, "Slick 'Ems, Glick 'Ems, Christmas Trees, and Cookie Cutters: Nuclear Language and How We Learned to Pat the Bomb," *Bulletin of the Atomic Scientists* 43 (June 1987): 19.

2. Paul Boyer, "From Activism to Apathy: The American People and Nuclear Weapons, 1963–1980," *Journal of American History* 70 (Mar. 1984): 821–44; Winkler, "The Nuclear Question," 148.

3. Winkler, *Modern America*, 207–8; *International Herald-Tribune*, 27–28 July 1985.

4. Jeffrey Richelson, "PD–59, NSDD–13 and the Reagan Strategic Modernization Program," *Journal of Strategic Studies* 6 (June 1983): 125, 129, 131; Michael Getler, "Administration's Nuclear War Policy Stance Still Murky," *Washington Post*, 10 Nov. 1982; Winkler, "The Nuclear Question," 137; "Arms Control at the Crossroads," *Newsweek*, 1 Oct. 1984, 28–30.

5. R. E. Lapp, "Atomic Bomb Explosions—Effects on an American City," *Bulletin of the Atomic Scientists* 4 (Feb. 1948): 50–53; note by the Secretary, Atomic Energy Commission, Atomic Attack on the City of Washington, 6 Oct. 1949, AEC 186/1, RG 326, U.S. Atomic Energy Commission, Secretariat Collection, Box 1222, Folder 384:51 Civil Defense, U.S. Department of Energy Archives; Hal Lindsey, *The Late Great Planet Earth* (Grand Rapids, Mich.: Zondervan, 1970), 162; "Thinking the Unthinkable," 36–38; *Eugene* [Oregon] *Register-Guard*, 5 Oct. 1982 and 15 Sept. 1983.

6. Michael Riordan, ed., *The Day After Midnight: The Effects of Nuclear War* (Palo Alto, Calif.: Cheshire, 1982), 29–67, 109, 119–25.

7. *Ambio*, Vol. 11, No. 2–3 (1982).

8. Jonathon Schell, *The Fate of the Earth* (New York: Alfred A. Knopf, 1982), 45.

9. Ibid., 154, 172.

10. "Books," *International Herald-Tribune*, 15 Apr. 1982.

11. Jonathon Schell, *The Abolition* (New York: Avon, 1984), 61, 134; Lord Zuckerman, "Nuclear Fantasies," *New York Review of Books*, 14 June 1984, 5.

12. Robert A. Heinlein, *Farnham's Freehold* (New York: New American Library, 1964); Robert C. O'Brien, *Z for Zachariah* (New York: Atheneum, 1975).

13. Bernard Malamud, *God's Grace* (New York: Farrar, Straus & Giroux, 1982), 3; Winkler, "American Hopes and Fears in the Nuclear Age," 37–38.

14. Whitley Strieber and James W. Kunetka, *Warday* (New York: Holt, Rinehart and Winston, 1984); William Prochnau, *Trinity's Child* (New York: G.P. Putnam's Sons, 1983); Dr. Seuss, *The Butter Battle Book* (New York: Random House, 1984); Herbert Kupferberg, "A Seussian Celebration," *Parade*, 26 Feb. 1984, 4.

15. Pink Floyd, "the final cut," Columbia Records, 1983.

16. Rosenblatt, "The Atomic Age," 46, 48.

17. Stephen Kneeshaw, "Hollywood and 'The Bomb,' " *OAH Newsletter*, May 1986, 10; H. Bruce Franklin, "Domesticating the Bomb: Nuclear Weapons in Fiction and Film of Judith Merril, Helen Clarkson, Kate Wilhelm, Carol Amen, and Lynne Littman," paper delivered at the American Studies Association Convention, Miami, Fla., 30 Oct. 1988, 10–11; Rosenblatt, "The Atomic Age," 48; Viewer's Guide to *Threads*, n.d.

18. "TV's Nuclear Nightmare," *Newsweek*, 21 Nov. 1983; Stephen Farber, "How a Nuclear Holocaust Was Staged for TV," *New York Times*, 13 Nov. 1983; *Eugene* [Oregon] *Register-Guard*, 15 Nov. 1983, 21 Nov. 1983, 23 Nov. 1983; Kneeshaw, "Hollywood and 'The Bomb,' " 10; Rosenblatt, "The Atomic Age," 48; Winkler, "American Hopes and Fears in the Nuclear Age," 38.

19. Stephanie Mansfield, "Citizen Woodward and Her Cause: Disarmament," *International Herald-Tribune*, 15–16 Sept. 1984.

20. Jonathon A. Leonard, "Danger: Nuclear War," *Harvard Magazine*, Nov.–Dec. 1980, 21–22; Bernard Lown quoted in Fox Butterfield, "Anatomy of the Nuclear Protest," *New York Times Magazine*, 11 July 1982, 17.

21. Howard Hiatt and Eric Chivian quoted in Leonard, "Danger: Nuclear War," 22, 24.

22. George Kistiakowsky quoted in Ibid., 22.

23. *New York Times*, 2 Mar. 1980; Leonard, "Danger: Nuclear War," 24.

24. Helen Caldicott quoted in Butterfield, "Anatomy of the Nuclear Protest," 17; *If You Love this Planet* (film); for an essay putting Caldicott's approach into historical prespective, see Paul Boyer, "A Historical View of Scare Tactics," *Bulletin of the Atomic Scientists* 42 (Jan. 1986): 17–19.

25. George Scialabba, "Books and Media: Nuclear War and Health," *Harvard Magazine*, Mar.–Apr. 1983, 73; H. Jack Geiger, "The Medical Effects on a City in the United States," and Robert Jay Lifton and Kai Erikson, "Survivors of Nuclear War: Psychological and Communal Breakdown," in International Physicians for the Prevention of Nuclear War, *Last Aid: The Medical Dimensions of Nuclear War*, edited by Eric Chivian, Susanna Chivian, Robert Jay Lifton, and John E. Mack (San Francisco: W.H. Freeman, 1982), 150, 289, 292.

26. George Kistiakowsky, "The Four Anniversaries," *Bulletin of the Atomic Scientists* 38 (Dec. 1982): 3.

27. Henry W. Kendall quoted in Butterfield, "Anatomy of the Nuclear Protest," 17; Winkler, "The Nuclear Question," 149.

28. Epstein, *Political Protest and Cultural Revolution*, 12–13.

29. "The Challenge of Peace: God's Promise and Our Response," in Jim Castelli, *The Bishops and the Bomb: Waging Peace in a Nuclear Age* (Garden City, N.Y.: Image Books, 1983), 191–92.

30. R. P. Turco, O. B. Toon, T. P. Ackerman, J. B. Pollack, and Carl Sagan, "Nuclear Winter: Global Consequences of Multiple Nuclear Explosions," *Science* 222 (23 Dec. 1983): 1,282–90; Carl Sagan, "The Nuclear Winter," *Parade*, 30 Oct. 1983, 4–5.

31. Summary of Conference Findings, The World After Nuclear War, Conference on the Long-Term Worldwide Biological Consequences of Nuclear War, 31 Oct.–1 Nov. 1983, Washington, D.C.; Sagan, "The Nuclear Winter," 5, 7.

32. Butterfield, "Anatomy of the Nuclear Protest," 32, 34.

33. "A Matter of Life and Death," *Newsweek*, 26 Apr. 1982, 21–22; Michael Mandelbaum, "Disarming Proposals," *New York Times Book Review*, 18 July 1982, 10; Physicians for Social Responsibility Memorandum from Wendy Silverman to Chapters and Chapters-in-Formation, Washington Office, Washington, D.C., 12 Aug. 1983; Edward M. Kennedy and Mark O. Hatfield, *Freeze! How You Can Help Prevent Nuclear War* (New York: Bantam, 1982), 132, 136, 169–70; *Hearings before the Committee on Foreign Relations, United States Senate*, 97th Congress, 2nd Session, 1982, 4–5.

34. Christopher M. Lehman, "Arms Control vs. the Freeze," in Steven E. Miller, ed., *The Nuclear Weapons Freeze and Arms Control* (Cambridge, Mass.: Ballinger, 1984), 65–71; *New York Times*, 13 June 1983; Russell Baker, "Nuclear Experts Need Dose of Doubt," *Eugene* [Oregon] *Register-Guard*, 13 Dec. 1982.

35. *PeacePAC 1982 Election Report* (brochure), PeacePAC, Washington D.C., n.d.

36. Sheila Tobias and Peter Goudinoff, "Understanding Star Wars," *Ms.*, Feb. 1986, 55; Paul Boyer, "Arms Race as Sitcom Plot," *Bulletin of the Atomic Scientists* 45 (June 1989): 7.

37. H. Bruce Franklin, *War Stars*, 202; Stephen Vaughn, "Spies, National Security, and the 'Inertia Projector': The Secret Service Films of Ronald Reagan," *American Quarterly* 39 (Fall 1987): 355, 368–69; Philip M. Boffey, William J. Broad, Leslie H. Gelb, Charles Mohr, and Holcomb B. Noble, *Claiming the Heavens: The New York Times Complete Guide to the Star Wars Debate* (New York: Times Books, 1988), 3–6; Richard Rhodes, "Peace Through Lasers?," *New York Times Book Review*, 20 Mar. 1988, 11.

38. Gregg Herken, "The Earthly Origins of Star Wars," *Bulletin of the Atomic Scientists* 43 (Oct. 1987): 20–21; Gregg Herken, "Science, Weapons, and Politics from Roosevelt to Reagan," address at Miami University, Oxford, Ohio, 19 Oct. 1987.

39. Ronald Reagan, "Address to the Nation on Defense and National Security," 23 Mar. 1983, in *Public Papers of the Presidents of the United States: Ronald*

Reagan, 1983, Book 1 (Washington, D.C.: U.S. Government Printing Office, 1984), 442–43; Herken, "The Earthly Origins of Star Wars," 27.

40. Jerry Adler, "The Star Warriors," *Newsweek*, 17 June 1985, 10–15; Union of Concerned Scientists, "Briefing Paper 5—The New Arms Race: Star Wars Weapons," n.d.; Zbigniew Brzezinski, Robert Jastrow, and Max. M. Kampelman, "Search for Security: The Case for the Strategic Defense Initiative," *International Herald-Tribune*, 28 Jan. 1985; Philip M. Boffey, "Proposed Space Defense Has Offensive Capability," *International Herald-Tribune*, 12 Mar. 1985.

41. Philip M. Boffey, "Many Questions Remain as 'Star Wars' Advances," *International Herald-Tribune*, 12 Mar. 1985.

42. Union of Concerned Scientists, *The Fallacy of Star Wars*, edited by John Tirman (New York: Vintage, 1984); Carl Sagan, "Star Wars: The Leaky Shield," *Parade*, 8 Dec. 1985, 16; Richard Garwin quoted in William D. Marbach, "Realistic Defense or Leap of Faith?," *Newsweek*, 17 June 1985, 16; Kosta Tsipis quoted in John E. Mack, "Action and Academia in the Nuclear Age," *Harvard Magazine*, Jan.–Feb. 1987, 29; *New York Times*, 15 Sept. 1985.

43. Mack, "Action and Academia," 29; Don Oberdorfer, "What the ABM Treaty Really Means Is . . . ," *Washington Post National Weekly Edition*, 4 Nov. 1985.

44. William Safire, "Acronym Sought," *New York Times Magazine*, 24 Feb. 1985, 9–10; William Safire, "New Name for 'Star Wars,' " *New York Times Magazine*, 24 Mar. 1985, 14, 16; Edward Tabor Linenthal, *Symbolic Defense: The Cultural Significance of the Strategic Defense Initiative* (Urbana and Chicago: Univ. of Illinois Press, 1989), 15–16, 108.

45. Art Buchwald, "The Sky's the Limit," *International Herald-Tribune*, 5–6 Jan. 1985; Kim McDonald, "Symbolic 'Clock' of Atomic Scientists Moved to 3 Minutes Before Midnight," *Chronicle of Higher Education*, 4 Jan. 1984, 2.

46. "A Start on Star Wars," *Newsweek*, 8 Feb. 1988, 30–32; Robert Scheer, "Edward Teller: His Livermore Laboratory under a Cloud of Scandal," *Cincinnati Enquirer*, 28 Aug. 1988; William J. Broad, "Beyond the Bomb. Turmoil in the Labs," *New York Times Magazine*, 9 Oct. 1988, 25, 72; Fred Kaplan, "Dream for SDI Fades," *Cincinnati Enquirer*, 8 Oct. 1989.

47. Bernard Lown quoted in Ellen Goodman, "Nobel Laureate Prescribes Test Ban," *Eugene* [Oregon] *Register Guard*, 4 Apr. 1986; Robert Jay Lifton quoted in Leonard, "Danger: Nuclear War," 25; Theodore Roethke cited in Robert J. Lifton and Nicholas Humphrey, eds., *In a Dark Time* (Cambridge, Mass.: Harvard Univ. Press, 1984), 140.

Epilogue

1. Alvin C. Graves cited by Daniel Lang, "Our Far-Flung Correspondents: Bombs Away!" *The New Yorker*, 10 May 1952, 76; Robert D. Benford and Lester R. Kurtz, "Performing the Nuclear Ceremony: The Arms Race as a Ritual," *Journal of Applied Behavioral Science* 23 (1987): 463–70.

2. Bernard Lown quoted in the *Eugene* [Oregon] *Register-Guard*, 9 Sept. 1982.

Bibliography

MANUSCRIPTS AND ARCHIVAL RECORDS

Committee for a Sane Nuclear Policy, Records. Swarthmore College Library, Swarthmore, Pa.

Eisenhower, Dwight D., Papers. Dwight D. Eisenhower Library, Abilene, Kans.

Federal Civil Defense Administration (FCDA), Records. Office of Civil and Defense Mobilization, Washington National Records Center, Suitland, Md.

Jackson, C. D., Papers. Dwight D. Eisenhower Library, Abilene, Kans.

Kennedy, John F., Papers. John F. Kennedy Library, Boston, Mass.

McNaughton, Frank, Papers. Harry S Truman Library, Independence, Mo.

Roosevelt, Franklin D., Papers. Franklin D. Roosevelt Library, Hyde Park, N.Y.

Stimson, Henry L., Papers. Manuscripts and Archives, Yale University Library, New Haven, Conn.

Truman, Harry S, Papers. Harry S Truman Library, Independence, Mo.

U.S. Department of Energy, Records. Coordination and Information Center, Reynolds Electrical & Engineering Co., Inc., Las Vegas, Nev.

——. U.S. Department of Energy Archives, Germantown, Md.

Yates, Sidney R., Papers. Harry S Truman Library, Independence, Mo.

PUBLIC DOCUMENTS

Calvert Cliffs Coordinating Committee v. AEC, 449 F.2d 1109, 1971.

Federal Civil Defense Administration. *Facts about Fallout.* n.d.

——. *What You Should Know about Radioactive Fallout.* n.d.

Government Indemnity: Hearings before the Joint Committee on Atomic Energy on Governmental Indemnity for Private Licensees and AEC Contractors against Reactor Hazards, 84th Cong., 2nd sess., 1956.

Hearings before the Committee on Foreign Relations, United States Senate, 97th Congress, 2nd Session, 1982.

House Military Operations Subcommittee of the Committee on Government Operations. *Civil Defense for National Survival, Hearings before a Subcommittee of the Committee on Government Operations,* 84th Congress, 2nd session, 1956.

248

Joint Committee on Atomic Energy. *Proposed Expanded Civilian Nuclear Power Program*. Washington, D.C.: U.S. Government Printing Office, 1958.

Office of Civil and Defense Mobilization. *How It Was Done*. n.d.

Reagan, Ronald. "Nuclear Energy Policy," 8 October 1981. *Weekly Compilation of Presidential Documents* 17 (12 Oct. 1981).

A Report on the International Control of Atomic Energy. Department of State Publication 2498. Washington, D.C.: U.S. Government Printing Office, 16 Mar. 1946.

Report of the President's Commission on the Accident at Three Mile Island—The Need for Change: The Legacy of TMI. Washington, D.C.: U.S. Government Printing Office, 1979.

Thirteenth Semiannual Report of the Atomic Energy Commission. Washington, D.C.: U.S. Government Printing Office, 1953.

U.S. Atomic Energy Commission. *Atomic Effects in the Nevada Test Site Region*. Jan. 1955.

——.*Civilian Nuclear Power . . . A Report to the President—1962*. 1962.

——. *In the Matter of J. Robert Oppenheimer: Texts of Principal Documents and Letters of Personnel Security Board, General Manager, Commissioners*. Washington, D.C.: U.S. Government Printing Office, 1954.

——. *In the Matter of J. Robert Oppenheimer: Transcript of Hearing before Personnel Security Board, Washington, D.C., April 12, 1954 through May 6, 1954*. Washington, D.C.: U.S. Government Printing Office, 1954.

——. *Plowshare*. 1966.

——. *Theoretical Possibilities and Consequences of Major Accidents in Large Nuclear Power Plants*. WASH–740, Mar. 1957. Washington, D.C.: U.S. States Atomic Energy Commission, 1957.

NEWSPAPERS

Cincinnati Enquirer
Eugene [Oregon] *Register-Guard*
International Herald-Tribune
Las Vegas Sun
Los Angeles Examiner
New York Times
Salt Lake City Tribune
San Francisco Examiner
Wall Street Journal
Washington Post

UNPUBLISHED MATERIALS

"Civil Defense: Are Federal, State, and Local Governments Prepared for Nuclear Attack?" Report LCD–76–464, by the comptroller general of the United States, 8 Aug. 1977.

Franklin, H. Bruce. "Domesticating the Bomb: Nuclear Weapons in Fiction and Film of Judith Merril, Helen Clarkson, Kate Wilhelm, Carol Amen, and Lynne Littman." Paper delivered at the American Studies Association Convention, Miami, Fla., 30 Oct. 1988.

Hales, Peter B. "The Atomic Sublime." Paper presented at the American Studies Association Convention, New York, 23 Nov. 1987.

Herken, Gregg. "Science, Weapons, and Politics from Roosevelt to Reagan." Address at Miami University, Oxford, Ohio, 19 Oct. 1987.

Mazuzan, George T., and Roger R. Trask. "An Outline History of Nuclear Regulation and Licensing, 1946–1979," Apr. 1979, 1–8, in Historical Office, Office of the Secretary, U.S. Nuclear Regulatory Commission, Washington, D.C.

Mazuzan, George T. and J. Samuel Walker. "The Safety Goal Issue in Historical Perspective," Dec. 1980, in Historical Office, Office of the Secretary, U.S. Nuclear Regulatory Commission, Washington, D.C.

PeacePAC 1982 Election Report (brochure). PeacePAC, 100 Maryland Avenue, N.E., Washington D.C. 20002, n.d.

Physicians for Social Responsibility Memorandum from Wendy Silverman to Chapters and Chapters-in-Formation, in Washington Office, 236 Massachusetts Avenue, N.E., Room 301, Washington, D.C. 20002, 12 Aug. 1983.

Schandler, Herbert Y. "U.S. Policy on the Use of Nuclear Weapons, 1945–1975." Congressional Research Service, 1975.

Summary of Conference Findings, The World After Nuclear War. Conference on the Long-Term Worldwide Biological Consequences of Nuclear War, 31 Oct.–1 Nov. 1983, 1735 New York Avenue, N.W., Suite 400, Washington, D.C. 20006.

Tyler, Lyon Gardiner, Jr. "Civil Defense: The Impact of the Planning Years, 1945–1950." Ph.D. dissertation, Duke University, 1967.

Union of Concerned Scientists. "Briefing Paper 5—The New Arms Race: Star Wars Weapons," n.d.

Walker, J. Samuel. Commentary in session on Problems of the Atomic Age. Organization of American Historians Convention, Cincinnati, Ohio, 9 Apr. 1983.

Yoshpe, Harry P. "Our Missing Shield: The U.S. Civil Defense Program in Historical Perspective, Final Report for Federal Emergency Management Agency." Contract No. DCPA 01–79-C–0294, Washington, D.C., Apr. 1981.

RECORDS

"Atomic Cafe: Radioactive Rock'n Roll, Blues, Country & Gospel." Rounder Records, n.d.

Lehrer, Tom. "An Evening Wasted with Tom Lehrer." Number 6199. Reprise Records, n.d.

——. "Songs by Tom Lehrer." Number 6216. Reprise Records, n.d.
——. "That Was the Year That Was." R–6179. Reprise Records, n.d.
Pink Floyd. "the final cut." Columbia Records, 1983.

BOOKS

Acheson, Dean. *Present at the Creation: My Years in the State Department.* New York: New American Library, 1969.

Ambio. Vol. 11, No. 2–3 (1982).

Ambrose, Stephen A. *Eisenhower*, Vol. 2, *The President.* New York: Simon & Schuster, 1984.

——. *Rise to Globalism: American Foreign Policy, 1938–1970.* Baltimore: Penguin, 1971.

Asimov, Isaac. *Opus 100.* Boston: Houghton Mifflin, 1969.

Ball, Howard. *Justice Downwind: America's Atomic Testing Program in the 1950s.* New York: Oxford Univ. Press, 1986.

Balogh, Brian. *Chain Reaction: Expert Debate & Public Participation in American Commercial Nuclear Power, 1945–1975.* New York: Cambridge Univ. Press, 1991.

Baxter, James Phinney, 3rd. *Scientists against Time.* Cambridge, Mass.: MIT Press, 1968.

Bealer, Nelson F. and Franklyn M. Branley. *Experiments with Atomics.* New York: Thomas Y. Crowell, 1965.

Bernstein, Barton J. and Allen J. Matusow, eds. *The Truman Administration: A Documentary History.* New York: Harper & Row, 1966.

Blumberg, Stanley A. and Gwinn Owens. *Energy and Conflict: The Life and Times of Edward Teller.* New York: G.P. Putnam's Sons, 1976.

Boffey, Philip M., William J. Broad, Leslie H. Gelb, Charles Mohr, and Holcomb B. Noble. *Claiming the Heavens: The New York Times Complete Guide to the Star Wars Debate.* New York: Times Books, 1988.

Bowles, Chester. *Promises to Keep: My Years in Public Life, 1941–1969.* New York: Harper & Row, 1971.

Boyer, Paul. *By the Bomb's Early Light: American Thought and Culture at the Dawn of the Atomic Age.* New York: Pantheon, 1985.

Bradley, David. *No Place to Hide, 1946/1984.* Hanover, N.H.: Univ. Press of New England, 1983.

Bremner, Robert H., Gary W. Reichard, and Richard J. Hopkins, eds. *American Choices: Social Dilemmas and Public Policy since 1960.* Columbus, Ohio: Ohio State Univ. Press, 1986.

Brenner, Michael J. *Nuclear Power and Non-Proliferation: The Remaking of U.S. Policy.* New York: Cambridge Univ. Press, 1981.

Brians, Paul. *Nuclear Holocausts: Atomic War in Fiction.* Kent, Ohio: Kent State Univ. Press, 1987.

Brodie, Bernard, ed. *The Absolute Weapon: Atomic Power and World Order.* New York: Harcourt, Brace, 1946.

Brodie, Bernard and Fawn. *From Crossbow to H-Bomb*. Revised and Enlarged Edition. Bloomington, Ind. Indiana Univ. Press, 1973.

Bundy, McGeorge. *Danger and Survival: Choices about the Bomb in the First Fifty Years*. New York: Random House, 1988.

Burdick, Eugene and Harvey Wheeler. *Fail-Safe*. New York: Dell, 1962.

Calder, Nigel. *Nuclear Nightmares: An Investigation into Possible Wars*. New York: Penguin, 1981.

Carrol, John M. and George C. Herring, eds. *Modern American Diplomacy*. Wilmington, Del.: Scholarly Resources, 1986.

Castelli, Jim. *The Bishops and the Bomb: Waging Peace in a Nuclear Age*. Garden City, N.Y.: Image Books, 1983.

Caufield, Catherine. *Multiple Exposures: Chronicles of the Radiation Age*. New York, Harper & Row, 1989.

Clarfield Gerard H. and William M. Wiecek. *Nuclear America: Military and Civilian Nuclear Power in the United States, 1940–1980*. New York: Harper & Row, 1984.

Conklin, Groff, ed. A *Treasury of Science Fiction*. New York: Bonanza, 1980.

Cox, Donald W. *America's New Policy Makers: The Scientists' Rise to Power*. New York: Chilton, 1964.

Dallin, Alexander, with Jonathon Harris and Grey Hodnett. *Diversity in International Communism: A Documentary Record, 1961–1963*. New York: Columbia Univ. Press, 1963.

Davis, Nuel Pharr. *Lawrence & Oppenheimer*. New York: Simon & Schuster, 1968.

Del Sesto, Steven L. *Science, Politics, and Controversy: Civilian Nuclear Power in the United States, 1946–1974*. Boulder, Colo.: Westview, 1979.

Dietz, David. *Atomic Energy in the Coming Era*. New York: Dodd, Mead & Co., 1945.

Dillard, Annie. *An American Childhood*. New York: Perennial Library 1988.

Divine, Robert A. *Blowing on the Wind: The Nuclear Test Ban Debate, 1954–1960*. New York: Oxford Univ. Press, 1978.

Donovan, Robert J. *Conflict and Crisis: The Presidency of Harry S Truman, 1945–1948*. New York: W.W. Norton, 1977.

Dowling, David. *Fictions of Nuclear Disaster*. Iowa City, Iowa: Univ. of Iowa Press, 1987.

Dyson, Freeman. *Disturbing the Universe*. New York: Harper & Row, 1979.

Einstein, Albert. *Out of My Later Years*. New York: Philosophical Library, 1950.

Eisenhower, Dwight D. *The White House Years: Mandate for Change, 1953–1956*. Garden City, N.Y.: Doubleday, 1963.

——. *The White House Years: Waging Peace, 1956–1961*. Garden City, N.Y.: Doubleday, 1965.

Epstein, Barbara. *Political Protest and Cultural Revolution: Nonviolent Direct Action in the 1970s and 1980s*. Berkeley and Los Angeles: Univ. of California Press, 1991.

Evans, Sara. *Born for Liberty: A History of Women in America*. New York: Free Press, 1989.

Fairlie, Henry. *The Kennedy Promise.* New York: Dell, 1973.

Faulkner, Peter, ed. *The Silent Bomb: A Guide to the Nuclear Energy Controversy.* New York: Vintage, 1977.

Ferrell, Robert H. *Harry S Truman and the Modern American Presidency.* Boston: Little, Brown, 1983.

——, ed. *Off the Record: The Private Papers of Harry S Truman.* New York: Harper & Row, 1980.

Fleming, Donald and Bernard Bailyn, eds. *Perspectives in American History,* Vol. 2, *1968.* Cambridge, Mass.: Charles Warren Center for Studies in American History, 1968.

Ford, Daniel. *The Cult of the Atom: The Secret Papers of the Atomic Energy Commission.* New York: Simon & Schuster, 1982.

Foreign Relations of the United States: 1945, Vol. 2, *General; Political and Economic Matters.* Washington, D.C.: U.S. Government Printing Office, 1967.

Foreign Relations of the United States: 1946, Vol. 1, *General; The United Nations.* Washington, D.C.: U.S. Government Printing Office, 1972.

Foreign Relations of the United States: 1947, Vol. 1, *General; The United Nations.* Washington, D.C.: U.S. Government Printing Office, 1973.

Foreign Relations of the United States: 1948, Vol. 1, Part 2, *General; The United Nations.* Washington, D.C.: U.S. Government Printing Office, 1976.

Foreign Relations of the United States: 1949, Vol. 1, *National Security Affairs, Foreign Economic Policy.* Washington, D.C.: U.S. Government Printing Office, 1976.

Foreign Relations of the United States: 1950, Vol. 1, *National Security Affairs; Foreign Economic Policy.* Washington, D.C.: U.S. Government Printing Office, 1977.

Foreign Relations of the United States: 1952–1954, Vol. 2, Part 1, *National Security Affairs.* Washington, D.C.: U.S. Government Printing Office, 1984.

[Forrestal, James]. *The Forrestal Diaries.* Edited by Walter Millis. New York: Viking, 1951.

Fradkin, Philip L. *Fallout: An American Nuclear Tragedy.* Tucson, Ariz.: Univ. of Arizona Press, 1989.

Frank, Pat. *Alas, Babylon.* New York: Bantam, 1970.

Franklin, H. Bruce. *War Stars: The Superweapon and the American Imagination.* New York: Oxford Univ. Press, 1988.

——, ed. *Countdown to Midnight.* New York: DAW Books, 1984.

Freeman, Ira M. *All About the Atom.* New York: Random House, 1955.

Freeman, Mae and Ira. *The Story of the Atom.* New York: Random House, 1960.

Friedan, Betty. *The Feminine Mystique.* New York: Dell, 1974.

Frost, Robert. *Complete Poems of Robert Frost, 1949.* New York: Henry Holt, 1949.

Fussell, Paul. *Thank God for the Atom Bomb and Other Essays.* New York: Summit Books, 1988.

Gaddis, John Lewis. *Strategies of Containment: A Critical Appraisal of Postwar American National Security Policy.* New York: Oxford Univ. Press, 1982.

Gallup, George H. *The Gallup Poll: Public Opinion, 1935–1971,* Vol. 1, *1935–1948.* New York: Random House, 1972.

——. *The Gallup Poll: Public Opinion, 1935–1971*, Vol. 2, *1949–1958*. New York: Random House, 1972.

Gamow, George. *Atomic Energy in Cosmic and Human Life*. New York: Macmillan, 1946.

George, Peter. *Dr. Strangelove or: How I Learned to Stop Worrying and Love the Bomb*. New York: Bantam, 1963.

Goldman, Eric F. *The Crucial Decade—and After: America, 1945–1960*. New York: Vintage, 1960.

Gosling, F. G., *The Manhattan Project: Science in the Second World War*. Washington, D.C.: History Division, Executive Secretariat, Office of Administration and Human Resources Management, U.S. Department of Energy, 1990.

Goulden, Joseph C. *The Best Years, 1945–1950*. New York: Atheneum, 1976.

Greenberg, Daniel S. *The Politics of Pure Science*. New York: New American Library, 1967.

Groves, Leslie. *Now It Can Be Told: The Story of the Manhattan Project*. New York: Harper & Row, 1962.

Haber, Heinz. *The Walt Disney Story of Our Friend the Atom*. New York: Simon & Schuster, 1956.

Hacker, Barton C. *The Dragon's Tail: Radiation Safety in the Manhattan Project, 1942–1946*. Berkeley and Los Angeles: Univ. of California Press, 1987.

Halberstam, David. *The Best and the Brightest*. New York: Random House, 1972.

Harbutt, Fraser J. *The Iron Curtain: Churchill, America, and the Origins of the Cold War*. New York: Oxford Univ. Press, 1986.

Harvard Nuclear Study Group (Albert Carnesale, Paul Doty, Stanley Hoffmann, Samuel P. Huntington, Joseph S. Nye, Jr., and Scott D. Sagan). *Living with Nuclear Weapons*. New York: Bantam, 1983.

Robert A. Heinlein, *Farnham's Freehold*. New York: New American Library, 1964.

[Heinlein, Robert A.] *The New Worlds of Robert A. Heinlein: Expanded Universe*. New York: Grosset & Dunlap, 1980.

Herken, Gregg. *Counsels of War*. New York: Alfred A. Knopf, 1985.

——. *The Winning Weapon: The Atomic Bomb in the Cold War, 1945–1950*. New York: Alfred A. Knopf, 1980.

Hersey, John. *Hiroshima*. New York: Bantam, 1966.

Hewlett, Richard G. and Oscar E. Anderson, Jr. *The New World, 1939–1946*, Vol. 1 of *A History of the United States Atomic Energy Commission*. Washington, D.C.: U.S. Atomic Energy Commission, 1972.

Hewlett Richard G. and Francis Duncan. *Atomic Shield, 1947–1952*, Vol. 2 of *A History of the United States Atomic Energy Commission*. Washington, D.C.: U.S. Atomic Energy Commission, 1972.

——. *Nuclear Navy, 1946–1962*. Chicago: Univ. of Chicago Press, 1974.

Hewlett, Richard G. and Jack M. Holl. *Atoms for Peace and War, 1953–1962: Eisenhower and the Atomic Energy Commission*. Berkeley and Los Angeles: Univ. of California Press, 1989.

Hilgartner, Stephen, Richard C. Bell, and Rory O'Connor. *Nukespeak: The Selling of Nuclear Technology in America*. New York: Penguin, 1983.

Hodgson, Godfrey. *America in Our Time*. Garden City, N.Y.: Doubleday, 1976.

Holl, Jack M., Roger M. Anders, and Alice L. Buck. *United States Civilian Nuclear Power Policy, 1954–1984: A Summary History*. Washington, D.C.: U.S. Department of Energy, Feb. 1986.

Hudson, Richard, and Ben Shahn. *Kuboyama and the Saga of the* Lucky Dragon. New York: Thomas Yoseloff, 1965.

Humphrey, Nicholas. *Consciousness Regained: Chapters in the Development of Mind*. New York: Oxford Univ. Press, 1984.

Ibuse, Masuku. *Black Rain*. Translated by John Bester. Palo Alto, Calif.: Kodansha International Ltd., 1969.

International Encyclopedia of the Social Sciences. Edited by David L. Sills. Vol. 15. New York: Macmillan and Free Press, 1968.

International Physicians for the Prevention of Nuclear War. *Last Aid: The Medical Dimensions of Nuclear War*. Edited by Eric Chivian, Susanna Chivian, Robert Jay Lifton, and John E. Mack. San Francisco: W.H. Freeman, 1982.

The Joan Baez Songbook. New York: Ryerson Music Publishers, 1964.

Jones, Vincent C. *Manhattan: The Army and the Atomic Bomb, Special Studies: United States Army in World War II*. Washington, D.C.: Center of Military History, U.S. Army, 1985.

Jungk, Robert. *Brighter Than a Thousand Suns: A Personal History of the Atomic Scientists*. New York: Harcourt Brace, 1958.

Kahn, Herman. *On Thermonuclear War*. Second Edition with Index. Princeton, N.J.: Princeton Univ. Press, 1961.

Kaplan, Fred. *The Wizards of Armageddon*. New York: Simon & Schuster, 1983.

Kaufmann, William W. *The McNamara Strategy*. New York: Harper & Row, 1964.

Kennedy, David M. and Michael E. Parrish, eds. *Power and Responsibility: Case Studies in American Leadership*. San Diego. Harcourt Brace Jovanovich, 1986.

Kennedy, Edward M. and Mark O. Hatfield. *Freeze! How You Can Help Prevent Nuclear War*. New York: Bantam, 1982.

Kennedy, Robert F. *Thirteen Days: A Memoir of the Cuban Missile Crisis*. New York: W.W. Norton, 1969.

Kerr, Thomas J. *Civil Defense in the U.S.: Bandaid for a Holocaust?* Boulder, Colo.: Westview, 1983.

Kevles, Daniel J. *The Physicists: The History of a Scientific Community in Modern America*. New York: Vintage, 1979.

Killian, James R., Jr. *Sputnik, Scientists and Eisenhower: A Memoir of the First Special Assistant to the President for Science and Technology*. Cambridge, Mass. MIT Press, 1977.

Kissinger, Henry A. *Nuclear Weapons and Foreign Policy*. New York: Harper & Brothers, 1957.

———. *White House Years*. Boston: Little, Brown, 1979.

Kolko, Gabriel. *The Politics of War: The World and United States Foreign Policy, 1943–1945*. New York: Random House, 1968.

Kovic, Ron. *Born on the Fourth of July*. New York: Pocket Books, 1977.

Lacey, Michael J., ed. *The Truman Presidency*. New York: Woodrow Wilson International Center for Scholars and Cambridge Univ. Press, 1989.

LaFollette, Marcel C. *Making Science Our Own: Public Images of Science, 1910–1955*. Chicago: Univ. of Chicago Press, 1990.

Lang, Daniel. *From Hiroshima to the Moon: Chronicles of Life in the Atomic Age*. New York: Simon & Schuster, 1959.

Lapp, Ralph E. *The Voyage of the* Lucky Dragon. New York: Harper & Brothers, 1958.

———. *The Weapons Culture*. New York: W.W. Norton, 1968.

Leaning, Jennifer. *Civil Defense in the Nuclear Age: What Purpose Does It Serve and What Survival Does It Promise?* Cambridge, Mass.: Physicians for Social Responsibility, 1982.

Lee, Stan. *Origins of Marvel Comics*. New York: Simon & Schuster, 1974.

Lehrer, Tom. *Too Many Songs by Tom Lehrer*. New York: Pantheon, 1981.

Lifton, Robert J. and Nicholas Humphrey, eds. *In a Dark Time*. Cambridge, Mass.: Harvard Univ. Press, 1984.

Lilienthal, David E. *Change, Hope, and the Bomb*. Princeton, N.J.: Princeton Univ. Press, 1963.

———. *The Journals of David E. Lilienthal*, Vol. 2, *The Atomic Energy Years, 1945–1950*. New York: Harper & Row, 1964.

Lindsey, Hal. *The Late Great Planet Earth*. Grand Rapids, Mich.: Zondervan, 1970.

Linenthal, Edward Tabor. *Symbolic Defense: The Cultural Significance of the Strategic Defense Initiative*. Urbana and Chicago: Univ. of Illinois Press, 1989.

Lingeman, Richard R. *Don't You Know There's a War On?: The American Home Front, 1941–1945*. New York: G.P. Putnam's Sons, 1970.

Lowell, Robert. *Selected Poems*. New York: Farrar, Straus & Giroux, 1976.

Malamud, Bernard. *God's Grace*. New York: Farrar, Straus & Giroux, 1982.

Manchester, William. *Goodbye, Darkness: A Memoir of the Pacific War*. Boston: Little, Brown, 1980.

Mandelbaum, Michael. *The Nuclear Question: The United States & Nuclear Weapons, 1946–1976*. New York: Cambridge Univ. Press, 1979.

———. *The Nuclear Revolution: International Politics before and after Hiroshima*. New York: Cambridge Univ. Press, 1981.

Marvel Tales Annual, #1, 1964. New York: Non-Pareil Publishing Corp., 1964.

Masters, Dexter and Katherine Way, eds. *One World or None*. New York: McGraw-Hill, 1946.

May, Elaine Tyler. *Homeward Bound: American Families in the Cold War Era*. New York: Basic Books, 1988.

Mazuzan, George T. and J. Samuel Walker. *Controlling the Atom: The Beginnings of Nuclear Regulation, 1946–1962*. Berkeley, Calif.: Univ. of California Press, 1984.

Miller, Steven E. *The Nuclear Weapons Freeze and Arms Control*. Cambridge, Mass.: Ballinger, 1984.

Miller, Walter M., Jr. *A Canticle for Leibowitz*. New York: Bantam, 1980.

Morison, Elting E. *Turmoil and Tradition: A Study of the Life and Times of Henry L. Stimson.* New York: Atheneum, 1964.

Murphy, Arthur W., ed. *The Nuclear Power Controversy.* Englewood Cliffs, N.J.: Prentice Hall, 1976.

Nathan, Otto and Heinz Norden, eds. *Einstein on Peace.* New York: Simon & Schuster, 1960.

Newhouse, John. *War and Peace in the Nuclear Age.* New York: Alfred A. Knopf, 1989.

Novick, Sheldon. *The Careless Atom.* Boston: Houghton Mifflin, 1969.

O'Brien, Robert C. *Z for Zachariah.* New York: Atheneum, 1975.

O'Brien, Tim. *The Nuclear Age.* New York: Alfred A. Knopf, 1985.

Olson, McKinley C. *Unacceptable Risk: The Nuclear Power Controversy.* New York: Bantam, 1976.

Parliamentary Debates (Hansard), Fifth Series—Volume 537, House of Commons, Official Report, Session 1954–55. London: Her Majesty's Stationary Office, 1955.

Parmet, Herbert S. *JFK: The Presidency of John F. Kennedy.* New York: Dial, 1983.

Patterson, Walter C. *Nuclear Power.* New York: Penguin, 1976.

Pearl, Carlton. *The Tenth Wonder: Atomic Energy.* Boston: Little, Brown, 1956.

Perrow, Charles. *Normal Accidents: Living with High-Risk Technologies.* New York: Basic Books, 1984.

Pfau, Richard. *No Sacrifice Too Great: The Life of Lewis L. Strauss.* Charlottesville, Va.: Univ. Press of Virginia, 1984.

Polenberg, Richard. *War and Society: The United States, 1941–1945.* New York: J.B. Lippincott, 1972.

Pringle, Peter and James Spigelman. *The Nuclear Barons.* New York: Holt, Rinehart and Winston, 1981.

Prochnau, William. *Trinity's Child.* New York: G.P. Putnam's Sons, 1983.

Public Papers of the Presidents of the United States : Dwight D. Eisenhower, 1953. Washington, D.C.: U.S. Government Printing Office, 1960.

Public Papers of the Presidents of the United States: Dwight D. Eisenhower, 1954. Washington, D.C.: U.S. Government Printing Office, 1960.

Public Papers of the Presidents of the United States: Harry S Truman, April 12 to December 31, 1945. Washington, D.C.: U.S. Government Printing Office, 1961.

Public Papers of the Presidents of the United States: Jimmy Carter, 1971. Book 2. Washington, D.C.: U.S. Government Printing Office, 1978.

Public Papers of the Presidents of the United States: John F. Kennedy, 1961. Washington, D.C.: U.S. Government Printing Office, 1962.

Public Papers of the Presidents of the United States : John F. Kennedy, 1962. Washington, D.C.: U.S. Government Printing Office, 1963.

Public Papers of the Presidents of the United States: John F. Kennedy, 1963. Washington, D.C.: U.S. Government Printing Office, 1964.

Public Papers of the Presidents of the United States: Ronald Reagan, 1983. Book 1. Washington, D.C.: U.S. Government Printing Office, 1984.

Pyle, Ernie. *Ernie's War: The Best of Ernie Pyle's World War II Dispatches*. Edited by David Nichols. New York: Simon & Schuster, 1986.

Reactor Safety Study: An Assessment of Risks in U.S. Commercial Nuclear Power Plants. WASH–1400 (NUREG–75/014), Oct. 1975. Springfield, Va.: U.S. Department of Commerce, National Technical Information Service, 1975.

Rhodes, Richard. *The Making of the Atomic Bomb*. New York: Simon & Schuster, 1986.

Riordan, Michael, ed. *The Day After Midnight: The Effects of Nuclear War*. Palo Alto, Calif.: Cheshire Books, 1982.

Rosenberg, Howard L. *Atomic Soldiers: American Victims of Nuclear Experiments*. Boston: Beacon, 1980.

Sargent, Pamela, ed. *Women of Wonder: Science Fiction Stories by Women about Women*. New York: Penguin, 1978.

Scheer, Robert. *With Enough Shovels: Reagan, Bush, and Nuclear War*. New York: Random House, 1982.

Schell, Jonathon. *The Abolition*. New York: Avon, 1984.

———. *The Fate of the Earth*. New York: Alfred A. Knopf, 1982.

Schlesinger, Arthur M., Jr. *A Thousand Days: John F. Kennedy in the White House*. Boston: Houghton Mifflin, 1965.

Seaborg, Glenn T. *Kennedy, Khrushchev, and the Test Ban*. Berkeley and Los Angeles: Univ. of California Press, 1983.

Seuss, Dr. *The Butter Battle Book*. New York: Random House, 1984.

Shaheen, Jack G., ed. *Nuclear War Films*. Carbondale and Edwardsville, Ill.: Southern Illinois Univ. Press, 1978.

Sherry, Michael S. *The Rise of American Air Power: The Creation of Armageddon*. New Haven, Conn.: Yale Univ. Press, 1987.

Sherwin, Martin J. *A World Destroyed: The Atomic Bomb and the Grand Alliance*. New York: Alfred A. Knopf, 1975.

Shute, Nevil. *On the Beach*. New York: William Morrow, 1957.

Silber, Norman Isaac. *Test and Protest: The Influence of Consumers Union*. New York: Holmes & Meier, 1983.

Sills, David L., C. P. Wolf, and Vivien B. Shelanski. *Accident at Three Mile Island: The Human Dimensions*. Boulder, Colo.: Westview, 1982.

Small, Fred. *Breaking from the Line: The Songs of Fred Small*. Cambridge, Mass.: Yellow Moon, 1986.

Smith, A. Merriman. *Thank You, Mr. President: A White House Notebook*. New York: Harper & Brothers, 1946.

Smith, Alice Kimball. *A Peril and a Hope: The Scientists' Movement in America, 1945–1947*. Chicago: Univ. of Chicago Press, 1965.

Smith, Alice Kimball and Charles Weiner, eds. *Robert Oppenheimer: Letters and Recollections*. Cambridge, Mass.: Harvard Univ. Press, 1980.

Smith, Gerard. *Doubletalk: The Story of the First Strategic Arms Limitation Talks*. Garden City, N.Y.: Doubleday, 1980.

Smoke, Richard. *National Security and the Nuclear Dilemma: An Introduction to the American Experience*. Reading, Mass.: Addison-Wesley, 1984.

Sorensen, Theodore C. *Kennedy*. New York: Harper & Row, 1965.

Stanford Arms Control Group. *International Arms Control: Issues and Agreements*,

Edited by Coit D. Blacker and Gloria Duffy. Stanford, Calif.: Stanford Univ. Press, 1984.

Stern, Philip M. *The Oppenheimer Case: Security on Trial*. New York: Harper & Row, 1969.

Stoff, Michael B., Jonathon F. Fanton, and R. Hal Williams, eds. *The Manhattan Project: A Documentary Introduction to the Atomic Age*. New York: McGraw-Hill, 1991.

Strauss, Lewis L. *Men and Decisions*. Garden City, N.Y.: Doubleday, 1962.

Strieber, Whitley and James W. Kunetka. *Warday*. New York: Holt, Rinehart and Winston, 1984.

Taylor, Maxwell D. *The Uncertain Trumpet*. New York: Harper & Brothers, 1960.

Teller, Edward. *The Legacy of Hiroshima*. Garden City, N.Y.: Doubleday, 1962.

Truman, Harry S. *Memoirs*, Vol. 1, *Year of Decisions*. Garden City, N.Y.: Doubleday, 1955.

——. *Memoirs*, Vol. 2, *Years of Trial and Hope*. Garden City, N.Y.: Doubleday, 1956.

Truman, Margaret. *Harry S Truman*. New York: William Morrow, 1973.

Tyler, Anne. *Dinner at the Homesick Restaurant*. New York: Berkeley Books, 1983.

Unforgettable Fire: Pictures Drawn by Atomic Bomb Survivors. Edited by the Japanese Broadcasting Corporation. New York: Pantheon, 1977.

Union of Concerned Scientists. *The Fallacy of Star Wars*. Edited by John Tirman. New York: Vintage, 1984.

The United States Strategic Bombing Survey: The Effects of Atomic Bombs on Hiroshima and Nagasaki. Washington: U.S. Government Printing Office, 1946.

Vonnegut, Kurt, Jr. *Player Piano*. New York: Bard, 1969.

——. *Slaughterhouse-Five*. New York: Dell, 1968.

Walker, J. Samuel. *Containing the Atom: Nuclear Regulation in a Changing Environment*. Berkeley: Univ. of California Press, 1992.

Walton, Richard J. *Cold War and Counterrevolution: The Foreign Policy of John F. Kennedy*. Baltimore: Penguin, 1973.

Weart, Spencer, R. *Nuclear Fear: A History of Images*. Cambridge, Mass.: Harvard Univ. Press, 1988.

Wells, H. G. *The World Set Free*. New York: E.P. Dutton, 1914.

Wigner, Eugene P., ed. *Who Speaks for Civil Defense?* New York: Charles Scribner's Sons, 1968.

Williams, Robert C. and Philip L. Cantelon, eds. *The American Atom: A Documentary History of Nuclear Policies from the Discovery of Fission to the Present, 1939–1984*. Philadelphia: Univ. of Pennsylvania Press, 1984.

Winkler, Allan M. *Modern America: The United States from World War II to the Present*. New York: Harper & Row, 1985.

Wittner, Lawrence S. *Cold War America: From Hiroshima to Watergate*. Expanded Edition. New York: Holt, Rinehart and Winston, 1978.

——. *Rebels against War: The American Peace Movement, 1941–1960*. New York: Columbia Univ. Press, 1969.

Wyden, Peter. *Day One: Before Hiroshima and After*. New York: Simon & Schuster, 1984.
York, Herbert F. *The Advisors: Oppenheimer, Teller, and the Superbomb*. Stanford, Calif.: Stanford Univ. Press, 1976.
———. *Making Weapons, Talking Peace: A Physicist's Odyssey from Hiroshima to Geneva*. New York: Basic Books, 1987.
———. *Race to Oblivion*. New York: Simon & Schuster, 1970.

ARTICLES

Abrash, Merritt. "Through Logic to Apocalypse: Science-Fiction Scenarios of Nuclear Deterrence Breakdown." *Science-Fiction Studies* 13 (July 1986).
Acheson, Dean. "Dean Acheson's Version of Robert Kennedy's Version of the Cuban Missile Affair: Homage to Plain Dumb Luck." *Esquire*, Feb. 1969.
Ackland, Len. "Dawn of the Atomic Age." *Chicago Tribune Magazine*, 28 Nov. 1982.
———. "Fighting for Time." *Chicago Tribune Magazine*, 11 Mar. 1984.
Adler, Jerry. "The Star Warriors." *Newsweek*, 17 June 1985.
"All Out Against Fallout." *Time*, 4 Aug. 1961.
Alsop, Stewart and Charles Bartlett. "In Time of Crisis." *The Saturday Evening Post*, 8 Dec. 1962.
Amrine, Michael. "The Issue of Fall-Out." *Current History*, Oct. 1957.
"Anatomic Bomb." *Life*, 3 Sept. 1945.
Anderson, Poul and F. N. Waldrop. "Tomorrow's Children." *A Treasury of Science Fiction*. Edited by Groff Conklin. New York: Bonanza, 1980.
"Are Nuclear Plants Winning Acceptance?" *Electrical World* 165 (24 Jan. 1966).
"Arms Control at the Crossroads." *Newsweek*, 1 Oct. 1984.
"At Elm & Main." *Time*, 30 Mar. 1953.
"The Atlantic Report: Science and Industry." *Atlantic Monthly* 213 (Mar. 1964).
"The Atom: Political Fallout." *Newsweek*, 17 June 1957.
"Atomic Blast at Yucca Flat." *Business Week*, 3 May 1952.
"Atomic Power: Cinderella Is Slipping Back into the Kitchen." *Science* 136 (20 Apr. 1962).
Augur, Tracy B. "The Dispersal of Cities—A Feasible Program." *Bulletin of the Atomic Scientists* 4 (Oct. 1948).
———. "The Dispersal of Cities as a Defense Measure." *Bulletin of the Atomic Scientists* 4 (May 1948).
Baker, Russell. "Nuclear Experts Need Dose of Doubt." *Eugene* [Oregon] *Register-Guard*, 13 Dec. 1982.
Baskin, Yvonne. "Too Intimate With the Atom." *New York Times Book Review*, 7 Jan. 1990.
Benedict, Ruth. "The Past and the Future." *The Nation*, 7 Dec. 1946.
Benford, Robert D. and Lester R. Kurtz. "Performing the Nuclear Ceremony: The Arms Race as a Ritual." *Journal of Applied Behavioral Science* 23 (1987).

Bernstein, Barton J. "The Challenges and Dangers of Nuclear Weapons: American Foreign Policy and Strategy, 1941–1961." *Foreign Service Journal*, Sept. 1978.

——. "The Quest for Security: American Foreign Policy and International Control of Atomic Energy, 1942–1946." *Journal of American History* 60 (Mar. 1974).

——. "Roosevelt, Truman, and the Atomic Bomb, 1941–1945: A Reinterpretation." *Political Science Quarterly* 90 (Spring 1975).

Bethe. H. A. "Can Air or Water Be Exploded?" *Bulletin of the Atomic Scientists* 1 (15 Mar. 1946).

Blank, Joseph P. "Atomic Power Comes of Age," *The Reader's Digest* 87 (Dec. 1965).

Bodansky, David and Fred H. Schmidt. "Safety Aspects of Nuclear Energy." *The Nuclear Power Controversy*. Edited by Arthur W. Murphy. Englewood Cliffs, N.J.: Prentice Hall, 1976.

Boffey, Philip M. "Many Questions Remain as 'Star Wars' Advances." *International Herald-Tribune*, 12 Mar. 1985.

——. "Proposed Space Defense Has Offensive Capability." *International Herald-Tribune*, 12 Mar. 1985.

"Bomb Scares." *Newsweek*, 28 June 1982, 73.

"Books." *International Herald-Tribune*, 15 Apr. 1982.

Boyer, Paul. "Arms Race as Sitcom Plot." *Bulletin of the Atomic Scientists* 45 (June 1989).

——. " "From Activism to Apathy: The American People and Nuclear Weapons, 1963–1980." *Journal of American History* 70 (Mar. 1984).

——. "A Historical View of Scare Tactics." *Bulletin of the Atomic Scientists* 42 (Jan. 1986).

——. "Physicians Confront the Apocalypse: The American Medical Profession and the Threat of Nuclear War." *JAMA—Journal of the American Medical Association* 254 (2 Aug. 1985).

——. " 'Some Sort of Peace': President Truman, the American People, and the Atomic Bomb." *The Truman Presidency*. Edited by Michael J. Lacey. New York: Woodrow Wilson International Center for Scholars and Cambridge Univ. Press, 1989.

Boyle, Robert H. "The Nukes Are in Hot Water." *Sports Illustrated*, 20 Jan. 1969.

Bracken, Paul. "Books." *International Herald-Tribune*, 19 Dec. 1984.

Broad, William J. "Beyond the Bomb: Turmoil in the Labs." *New York Times Magazine*, 9 Oct. 1988.

——. "The Men Who Made the Sun Rise." *New York Times Book Review*, 8 Feb. 1987.

——. "A Mountain of Trouble." *New York Times Magazine*, 18 Nov. 1990.

——. "Rewriting the History of the H-Bomb." *Science* 218 (19 Nov. 1982).

Brodie, Bernard. "Implications for Military Strategy." *The Absolute Weapon: Atomic Power and World Order*. Edited by Bernard Brodie. New York: Harcourt Brace, 1946.

——. "Strategy." *International Encyclopedia of the Social Sciences.* Edited by David L. Sills. New York: Macmillan and Free Press, 1968.

——. "War in the Atomic Age." *The Absolute Weapon: Atomic Power and World Order.* Edited by Bernard Brodie. New York: Harcourt Brace, 1946.

Brown, JoAnne. "'A Is for *Atom, B* Is for *Bomb*': Civil Defense in American Public Education, 1948–1963." *Journal of American History* 75 (June 1988).

Brzezinski, Zbigniew, Robert Jastrow, and Max M. Kampelman. "Search for Security: The Case for the Strategic Defense Initiative." *International Herald-Tribune,* 28 Jan. 1985.

Buchwald, Art. "The Sky's the Limit." *International Herald-Tribune,* 5–6 Jan. 1985.

"Building: Shelter Skelter." *Time,* 1 Sept. 1961.

Bundy, McGeorge. "The Missed Chance to Stop the H-Bomb." *New York Review of Books* 29 (13 May 1982).

Butterfield, Fox. "Anatomy of the Nuclear Protest." *New York Times Magazine,* 11 July 1982.

California Assembly, Committee on Resources, Land Use and Energy. "Reactor Safety." *The Silent Bomb: A Guide to the Nuclear Energy Controversy.* Edited by Peter Faulkner. New York: Vintage, 1977.

"Chernobyl's Cancer Toll Will Surpass Million, a Berkeley Physicist Predicts." *Chronicle of Higher Education,* 17 Sept. 1986.

"Civil Defense: Best Defense? Prayer." *Time,* 27 June 1955.

"Civil Defense: Coffins or Shields?" *Time,* 2 Feb. 1962.

"Civil Defense: A Place to Hide." *Time,* 18 December 1950.

"Civil Defense: The Sheltered Life." *Time,* 20 Oct. 1961.

"Civil Defense: So Much to Be Done." *Newsweek,* 27 June 1955.

"Civil Defense: Survival (Contd.)," *Time,* 3 Nov. 1961.

Coates, Peter. "Project Chariot: Alaskan Roots of Environmentalism." *Alaska History* (Fall 1989).

Cohn, Carol. "Slick 'Ems, Glick 'Ems, Christmas Trees, and Cookie Cutters: Nuclear Language and How We Learned to Pat the Bomb." *Bulletin of the Atomic Scientists* 43 (June 1987).

Cook, Blanche Wiesen. "The Legacy of Atomic Testing." *New York Times Book Review,* 1 Aug. 1982.

Cook, James. "Nuclear Follies." *Forbes,* 11 Feb. 1985.

[Cousins, Norman]. "Modern Man Is Obsolete." *The Saturday Review of Literature,* 18 Aug. 1945.

Cushing, Richard J. "A Spiritual Approach to the Atomic Age." *Bulletin of the Atomic Scientists* 4 (July 1948).

de Seversky, Alexander P. "Atomic Bomb Hysteria." *The Reader's Digest,* Feb. 1946.

Dean, Gordon. "Today: Atoms *Work* for You." *Parade,* 24 Feb. 1952.

"The Debate Goes On." *Newsweek,* 4 Dec. 1961.

"Does Civil Defense Make Sense?" *Newsweek,* 26 Apr. 1982.

Dreher, Carl. "Atomic Power: The Fear and the Promise." *The Nation* 198 (23 Mar. 1964).

Dulles, John Foster. "The Evolution of Foreign Policy." *Department of State Bulletin* 30 (25 Jan. 1954).

DuShane, Graham. "Loaded Dice." *Science* 125 (17 May 1957).

"Editorial." *The Christian Century*, 15 Aug. 1945.

Einstein, Albert, as told to Raymond Swing. "Einstein on the Atomic Bomb." *Atlantic Monthly*, Nov. 1945.

Emmons, Steve. "Sick Jokes: Coping With the Horror." *Los Angeles Times*, 30 May 1986.

Erskine, Hazel Gaudet. "The Polls: Atomic Weapons and Nuclear Energy." *Public Opinion Quarterly* 27 (Summer 1963).

"Fallout Cover-Up." *Newsweek*, 30 Apr. 1979.

"Fallout Shelters." *Life*, 15 Sept. 1961.

"Fallout Shelters: Low-Key Plan." *Newsweek*, 8 Jan. 1962.

"Fallout Shelters: Making a Comeback." *Newsweek*, 22 Feb. 1982.

Farber, Stephen. "How a Nuclear Holocaust Was Staged for TV." *New York Times*, 13 Nov. 1983.

Fitzsimons, Neal. "Brief History of Civil Defense." *Who Speaks for Civil Defense?* Edited by Eugene P. Wigner. New York: Charles Scribner's Sons, 1968.

Flynn, Cynthia Bullock. "Reactions of Local Residents to the Accident at Three Mile Island." *Accident at Three Mile Island: The Human Dimensions*. Edited by David L. Sills, C. P. Wolf, and Vivien B. Shelanski. Boulder, Colo.: Westview, 1982.

Franklin, H. Bruce. "Fatal Fiction: A Weapon to End All Wars." *Bulletin of the Atomic Scientists* 45 (Nov. 1989).

———. "Nuclear War and Science Fiction." *Countdown to Midnight*, Edited by H. Bruce Franklin. New York: DAW Books, 1984.

Friedberg, Aaron L. "A History of the U.S. Strategic 'Doctrine'—1945 to 1980." *Journal of Strategic Studies* 3 (Dec. 1980).

"A Frightening Message for a Thanksgiving Issue." *Good Housekeeping*, Nov. 1958.

Fuller, John G. "We Almost Lost Detroit." *The Silent Bomb: A Guide to the Nuclear Energy Controversy*. Edited by Peter Faulkner. New York: Vintage, 1977.

"Gags Away." *The New Yorker*, 18 Aug. 1945.

Gates, David. "Thinking About the Apocalypse." *Newsweek*, 2 Mar. 1987.

Geiger, H. Jack. "The Medical Effects on a City in the United States." International Physicians for the Prevention of Nuclear War. *Last Aid: The Medical Dimensions of Nuclear War*. Edited by Eric Chivian, Susanna Chivian, Robert Jay Lifton and John E. Mack. San Francisco: W. H. Freeman, 1982.

"General Taylor—Questions U.S. Army Policy." *U.S. News & World Report*, 6 July 1959.

Getler, Michael. "Administration's Nuclear War Policy Stance Still Murky." *Washington Post*, 10 Nov. 1982.

Gleick, James. "After the Bomb, a Mushroom Cloud of Metaphors." *New York Times Book Review*, 21 May 1989.

Goodman, Ellen. "A Big Hand, Folks, for the Nuclear Follies." *Eugene* [Oregon] *Register-Guard*, 10 Dec. 1983.

——. "Civil Defense Is Really Military Madness." *Eugene* [Oregon] *Register-Guard*, 2 Feb. 1982.

——. "Nobel Laureate Prescribes Test Ban." *Eugene* [Oregon] *Register-Guard*, 4 Apr. 1986.

"Greasewood Fires and Man's Most Terrible Weapon." *Newsweek*, 30 Mar. 1953.

Green, Harold P. "The Peculiar Politics of Nuclear Power." *Bulletin of the Atomic Scientists* 38 (Dec. 1982).

Grendon, Edward. "The Figure." *A Treasury of Science Fiction*. Edited by Groff Conklin. New York: Bonanza, 1980.

Griffith, Robert. "Dwight D. Eisenhower and the Corporate Commonwealth." *American Historical Review* 87 (Feb. 1982).

Gruber, Carol S. "Manhattan Project Maverick: The Case of Leo Szilard." *Prologue* 15 (Summer 1983).

"H. C. Urey on State Dept. Report." *Bulletin of the Atomic Scientists* 1 (Apr. 1, 1946).

Harper, John L. "Henry Stimson and the Origin of America's Attachment to Atomic Weapons." *SAIS Review* 5 (Summer–Fall 1985).

Heimann, Fritz F. "How Can We Get the Nuclear Job Done?" *The Nuclear Power Controversy*. Edited by Arthur W. Murphy. Englewood Cliffs, N.J.: Prentice Hall, 1976.

Heinlein, Robert A. "The Last Days of the United States." [Heinlein, Robert A.] *The New Worlds of Robert A. Heinlein: Expanded Universe*. New York: Grosset & Dunlap, 1980.

Herken, Gregg. "The Earthly Origins of Star Wars." *Bulletin of the Atomic Scientists* 43 (Oct. 1987).

Hirsch, Daniel and William G. Mathews. "The H-Bomb: Who Really Gave Away the Secret?" *Bulletin of the Atomic Scientists* 46 (Jan.–Feb. 1990).

Honicker, Clifford T. "The Hidden Files: America's Radiation Victims." *New York Times Magazine*, 19 Nov. 1989.

"How Fatal Is the Fall-Out?" *Time*, 22 Nov. 1954.

Humphrey, Nicholas. "The Bronowski Memorial Lecture: 'Four Minutes to Midnight.' " *The Listener*, 29 Oct. 1981.

"Hydrogen Bomb Deviser: Dr. Edward Teller." *New York Times*, 27 June 1957.

Kaplan, Fred. "Dream for SDI Fades." *Cincinnati Enquirer*, 8 Oct. 1989.

——. "The Soviet Civil Defense Myth, Part 2." *Bulletin of the Atomic Scientists* 34 (Apr. 1978).

——. "Strategic Thinkers." *Bulletin of the Atomic Scientists* 38 (Dec. 1982).

Ketchum, Richard M. "An Uncivil Defense." *The Reader's Digest*, Feb. 1983.

Keyerleber, Joseph. "On the Beach." *Nuclear War Films*. Edited by Jack G. Shaheen. Carbondale and Edwardsville, Ill.: Southern Illinois Univ. Press, 1978.

Kistiakowsky, George. "The Four Anniversaries." *Bulletin of the Atomic Scientists* 38 (Dec. 1982).

Kneeshaw, Stephen. "Hollywood and 'The Bomb.' " *OAH Newsletter*, May 1986.

Kopp, Carolyn. "The Origins of the American Scientific Debate over Fallout Hazards." *Social Studies of Science* 9 (1979).

Kraus, Sidney, Reuben Mehling, and Elaine El-Assal. "Mass Media and the Fallout Controversy." *Public Opinion Quarterly* 27 (Summer 1963).

Kupferberg, Herbert. "A Seussian Celebration." *Parade*, 26 Feb. 1984.

Lang, Daniel. "Our Far-Flung Correspondents: Bombs Away!" *The New Yorker*, 10 May 1952, 76.

Lapp, R[alph] E. "Atomic Bomb Explosions—Effects on an American City." *Bulletin of the Atomic Scientists* 4 (Feb. 1948).

——. "Civil Defense Faces New Peril." *Bulletin of the Atomic Scientists* 10 (Nov. 1954).

——. "The Strategy of Civil Defense." *Bulletin of the Atomic Scientists* 6 (Aug.–Sept. 1950).

Laurence, William L. "Is Atomic Energy the Key to Our Dreams?" *The Saturday Evening Post*, 13 Apr. 1946.

——. "Nagasaki Was the Climax of the New Mexico Test." *Life*, 24 Sept. 1945.

Lee, Stan, and S. Ditko. "Spider-Man!" *Marvel Tales Annual, #1, 1964*. New York: Non-Pareil Publishing Corp., 1964.

Lee, Stan, and J. Kirby. "The Hulk." *Marvel Tales Annual, #1, 1964*. New York: Non-Pareil Publishing Corp., 1964.

Lehman, Christopher M. "Arms Control vs. the Freeze." *The Nuclear Weapons Freeze and Arms Control*. Edited by Steven E. Miller. Cambridge, Mass.: Ballinger, 1984.

Leonard, Jonathon A. "Danger: Nuclear War." *Harvard Magazine*, Nov.–Dec. 1980.

"Let's Prepare . . . Shelters." *Life*, 13 Oct. 1961.

Lewis, Anthony. "Kennedy Showed That He Could Learn." *Eugene* [Oregon] *Register-Guard*, 16 Nov. 1983.

——. "Lessons of Chernobyl Lost on Leaders." *Eugene* [Oregon] *Register-Guard*, 30 May 1986.

Lewis, E. B. "Leukemia and Ionizing Radiation." *Science* 125 (17 May 1957).

Lifton, Robert Jay and Kai Erikson. "Survivors of Nuclear War: Psychological and Communal Breakdown." International Physicians for the Prevention of Nuclear War. *Last Aid: The Medical Dimensions of Nuclear War*. Edited by Eric Chivian, Susanna Chivian, Robert Jay Lifton and John E. Mack. San Francisco: W. H. Freeman, 1982.

L'il Abner. Script property of Tams-Witmark Music Library, Inc., 757 Third Avenue, New York, New York 10017.

Lilienthal, David E. "Democracy and the Atom." *NEA Journal*, Feb. 1948.

Linden, George W. "Dr. Strangelove or: How I Learned to Stop Worrying and Love the Bomb." *Nuclear War Films*. Edited by Jack G. Shaheen. Carbondale and Edwardsville, Ill.: Southern Illinois Univ. Press, 1978.

"Living with the Bomb: The First Generation of the Atomic Age." *Newsweek*, 29 July 1985.

"Love Thy Neighbor." *Newsweek*, 9 Oct. 1961.

Luft, Joseph and W. M. Wheeler. "Reaction to John Hersey's 'Hiroshima.'" *Journal of Social Psychology* 28 (Aug. 1948).

Mack, John E. "Action and Academia in the Nuclear Age." *Harvard Magazine*,
    Jan.–Feb. 1987.
———. " 'But what about the Russians?': A Psychiatrist Looks at American Nu-
    clear-Arms Paranoia." *Harvard Magazine*, Mar.–Apr. 1982.
[Macmillan, Harold]. "Introduction by Harold Macmillan." Robert F. Kennedy.
    Thirteen Days: A Memoir of the Cuban Missile Crisis. New York: W. W.
    Norton, 1969.
Malmsheimer, Lorna. "Three Mile Island: Fact, Frame and Fiction." *American
    Quarterly* 38 (Spring 1986).
Mandelbaum, Michael. "Disarming Proposals." *New York Times Book Review*,
    18 July 1982.
Mansfield, Stephanie. "Citizen Woodward and Her Cause: Disarmament." *In-
    ternational Herald-Tribune*, 15–16 Sept. 1984.
Marbach, William D. "Realistic Defense or Leap of Faith?" *Newsweek*, 17 June
    1985.
Mariotte, Michael. "Sudden Burst of Energy." *Nuclear Times*, July–Aug. 1986.
Marshak, J., E. Teller, and L. R. Klein. "Dispersal of Cities and Industries."
    Bulletin of the Atomic Scientists 1 (15 Apr. 1946).
"A Matter of Life and Death." *Newsweek* 26 Apr. 1982.
Mauck, Elwyn A. "History of Civil Defense in the United States." *Bulletin of
    the Atomic Scientists* 6 (Aug.–Sept. 1950).
Maynard, Joyce. "The Story of a Town." *New York Times Magazine*, 11 May
    1986.
Mazuzan, George T. "American Nuclear Policy." *Modern American Diplomacy*.
    Edited by John M. Carroll and George C. Herring. Wilmington, Del.:
    Scholarly Resources, 1986.
———. "Conflict of Interest: Promoting and Regulating the Infant Nuclear Power
    Industry, 1954–1956." *The Historian* 44 (Nov. 1981).
McDonald, Kim. "Symbolic 'Clock' of Atomic Scientists Moved to 3 Minutes
    Before Midnight." *Chronicle of Higher Education*, 4 Jan. 1984.
Meade, Mary E. "What Programs of Civil Defense Are Needed in Our Schools?"
    Bulletin of the National Association of Secondary-School Principals 36 (Apr.
    1952).
"Meltdown." *Newsweek*, 12 May 1986.
"A Meltdown for Nuclear Power." *Business Week*, 30 Jan. 1984.
Merril, Judith. "That Only a Mother." *Women of Wonder: Science Fiction Stories
    by Women about Women*. Edited by Pamela Sargent. New York: Penguin,
    1978.
"A Message to You from the President." *Life*, 15 Sept. 1961.
Messer, Robert L. "New Evidence on Truman's Decision." *Bulletin of the Atomic
    Scientists* 41 (Aug. 1985).
Metzner, Charles A. and Julia B. Kessler. "What Are the People Thinking?"
    Bulletin of the Atomic Scientists 7 (Nov. 1951).
Michael, Donald N. "Civilian Behavior under Atomic Bombardment." *Bulletin
    of the Atomic Scientists* 11 (May 1955).
"The Milk We Drink." *Consumer Reports* 24 (Mar. 1959).
Miller, Leslie J. "Children's Hot Cold War." *New York Times*, 12 June 1982.

Miller, Merle. "The Atomic Scientist in Politics." *Bulletin of the Atomic Scientists* 3 (Sept. 1947).

"More Fallout from Chernobyl." *Time*, 19 May 1986.

Morland, Howard. "The H-bomb Secret," *The Progressive*, Nov. 1979.

"National Defense: Ducking for Cover." *Newsweek*, 30 July 1956.

Newman, James. "Books: Two Discussions of Thermonuclear War." *Scientific American* 204 (Mar. 1961).

Nitze, Paul H. "Atoms, Strategy and Policy." *Foreign Affairs* 34 (Jan. 1956).

"Nuclear Accident." *Newsweek*, 9 Apr. 1979.

"A Nuclear Nightmare." *Time*, 9 Apr. 1979.

"Nuclear Power and the Community: Familiarity Breeds Confidence." *Nuclear News* 9 (May 1966).

"Nuclear Reactor Safety—the APS Submits Its Report." *Physics Today* 28 (July 1975).

Oberdorfer, Don. "What the ABM Treaty Really Means Is. . . . " *Washington Post National Weekly Edition*, 4 Nov. 1985.

Olum, Paul. "Why Did We Work to Build Such a Terrible Thing?" *The Oregonian*, 4 Aug. 1985.

"The Only Real Defense," *Bulletin of the Atomic Scientists* 7 (Sept. 1951).

Oppenheimer, J. Robert. "Atomic Weapons and American Policy." *Foreign Affairs* 31 (July 1953).

——. "Physics in the Contemporary World." *Bulletin of the Atomic Scientists* 4 (Mar. 1948).

"The Philosophers' Stone." *Time*, 15 Aug. 1955.

Pittman, Steuart L. "Government and Civil Defense." *Who Speaks for Civil Defense?* Edited by Eugene P. Wigner. New York: Charles Scribner's Sons, 1968.

Poore, Charles. " 'The Most Spectacular Explosion in the Time of Man.' " *New York Times Book Review*, 10 Nov. 1946.

"Pulling the Nuclear Plug." *Time*, 13 Feb. 1984.

Quammen, David. "The Conscience of the Young Scientist." *Esquire*, Dec. 1985.

Rabinowitch, Eugene. "Five Years After." *Bulletin of the Atomic Scientists* 7 (Jan. 1951).

"Radiation: 'Glimpse Into Hell.' " *Newsweek*, 13 Jan. 1958.

"Rasmussen Issues Revised Odds on a Nuclear Catastrophe." *Science* 190 (14 Nov. 1975).

Rasmussen, Norman C. "The Safety Study and Its Feedback." *Bulletin of the Atomic Scientists* 31 (Sept. 1975).

"Religion: Gun Thy Neighbor?" *Time*, 18 Aug. 1961.

Rhodes, Richard. "Peace Through Lasers?" *New York Times Book Review*, 20 Mar. 1988.

Richelson, Jeffrey. "PD–59, NSDD–13 and the Reagan Strategic Modernization Program." *Journal of Strategic Studies* 6 (June 1983).

Roback, Herbert. "Civil Defense and National Defense." *Who Speaks for Civil Defense?* Edited by Eugene P. Wigner. New York: Charles Scribner's Sons, 1968.

Rosenberg, David Alan. "American Atomic Strategy and the Hydrogen Bomb Decision." *Journal of American History* 66 (June 1979).

——. "The Origins of Overkill: Nuclear Weapons and American Strategy, 1945–1960." *International Security* 7 (Spring 1983).

——. " 'A Smoking Radiating Ruin at the End of Two Hours': Documents on American Plans for Nuclear War with the Soviet Union, 1954–1955." *International Security* 6 (Winter 1981–82).

Rosenblatt, Roger. "The Atomic Age." *Time*, 29 July 1985.

Russell, Bertrand. "The Danger to Mankind." *Bulletin of the Atomic Scientists* 10 (Jan. 1954).

Safire, William. "Acronym Sought." *New York Times Magazine*, 24 Feb. 1985.

——. "New Name for 'Star Wars.' " *New York Times Magazine*, 24 Mar. 1985.

Sagan, Carl. "The Nuclear Winter." *Parade*, 30 Oct. 1983.

——. "Star Wars: The Leaky Shield." *Parade*, 8 Dec. 1985.

Sagan, Scott D. "SIOP–62: The Nuclear War Plan Briefing to President Kennedy." *International Security* 12 (Summer 1987).

"SALT II: Sealed with a Kiss." *Newsweek*, 2 July 1979.

Scheer, Robert. "Edward Teller: His Livermore Laboratory under a Cloud of Scandal." *Cincinnati Enquirer*, 28 Aug. 1988.

Schrag, Philip G. "Seeing Ground Zero in Nevada." *New York Times*, 12 Mar. 1989.

Scialabba, George. "Books and Media: Nuclear War and Health." *Harvard Magazine*, Mar.–Apr. 1983.

Shaw, George Bernard. " 'What Does Atom Bomb Cost?' Shaw Asks—and Answers." *Washington Post*, 19 Aug. 1945.

"Shelters: Jitters Ease, Debate Goes On." *Newsweek*, 15 Jan. 1962.

Sherry, Michael. "The Slide to Total Air War." *The New Republic*, 16 Dec. 1981.

Sherwin, Martin J. "How Well They Meant." *Bulletin of the Atomic Scientists* 41 (Aug. 1985).

Sherwood, Robert E. "Please Don't Frighten Us." *Atlantic Monthly*, Feb. 1949.

Simpson, Mary M. "A Long Hard Look at Civil Defense." *Bulletin of the Atomic Scientists* 12 (Nov. 1956).

Slater, Leonard. "Atomic Tests: The World Blew Up." *Newsweek*, 19 Feb. 1951.

Smith, Alice Kimball. "Scientists and Public Issues." *Bulletin of the Atomic Scientists* 38 (Dec. 1982).

Smith, R. Jeffrey. "Scientists Implicated in Atom Test Deception." *Science* 218 (5 Nov. 1982).

"A Spare Room Fallout Shelter." *Life*, 25 Jan. 1960.

Spencer, Stephen M. "Fallout: The Silent Killer, Part One." *The Saturday Evening Post*, 29 Aug. 1959.

"A Start on Star Wars." *Newsweek*, 8 Feb. 1988.

Stimson, Henry L. "The Decision to Use the Atomic Bomb." *Harper's Magazine*, Feb. 1947.

Stone, I. F. "Atomic Pie in the Sky." *The Nation*, 6 Apr. 1946.

Storm, Grace. "Civil Defense Film for Schools." *Elementary School Journal* 52 (Sept. 12, 1952).

Strauss, Lewis. "My Faith in the Atomic Future." *The Reader's Digest*, Aug. 1955.

"Survival: Are Shelters the Answer?" *Newsweek*, 6 Nov. 1961.

Swerdlow, Amy. "Ladies Day at the Capitol: Women Strike for Peace Versus HUAC." *Feminist Studies* 8 (Fall 1982).

Szilard, Leo. "Reminiscences." *Perspectives in American History*, Vol. 2, 1968. Edited by Donald Fleming and Bernard Bailyn. Cambridge, Mass.: Charles Warren Center for Studies in American History, 1968.

"The Talk of the Town." *The New Yorker*, 21 Jan. 1985.

"Tatsue Urata's Story." *We of Nagasaki: The Story of Survivors in an Atomic Wasteland*. Edited by Takashi Nagai and translated by Ichiro Shirato and Herbert B. L. Silverman. New York: Duell, Sloan and Pearce, 1951.

Teller, Edward. "How Dangerous Are Atomic Weapons?" *Bulletin of the Atomic Scientists* 3 (Feb. 1947).

———. "To Sagittarius." *Bulletin of the Atomic Scientists* 7 (Jan. 1951).

———. "The State Dep't Report—'A Ray of Hope.' " *Bulletin of the Atomic Scientists* 1 (Apr. 1, 1946).

———. "The Work of Many People." *Science* 121 (25 Feb. 1955).

"Testimony Links Deaths to 'Beautiful Mushroom.' " *Eugene* [Oregon] *Register-Guard*, 29 Sept. 1982.

"Thinking the Unthinkable." *Newsweek*, 5 Oct. 1981.

Tobias, Sheila and Peter Goudinoff. "Understanding Star Wars." *Ms.*, Feb. 1986, 55.

Tucker, Robert B. "From Doom-and-Gloom to 'The Coming Boom.' " *United: The Magazine of the Friendly Skies*, July 1983.

Turco, R. P., O. B. Toon, T. P. Ackerman, J. B. Pollack, and Carl Sagan. "Nuclear Winter: Global Consequences of Multiple Nuclear Explosions." *Science* 222 (23 Dec. 1983).

"TV's Nuclear Nightmare." *Newsweek*, 21 Nov. 1983.

Vaughn, Stephen. "Spies, National Security, and the 'Inertia Projector': The Secret Service Films of Ronald Reagan." *American Quarterly* 39 (Fall 1987).

Walker, J. Samuel. "The Controversy Over Radiation Safety: A Historical Overview." *JAMA—Journal of the American Medical Association* 262 (4 Aug. 1989).

———. "Nuclear Power and the Environment: The Atomic Energy Commission and Thermal Pollution, 1965–1971." *Technology and Culture* 30 (Oct. 1989).

———. "Writing the History of Nuclear Energy: The State of the Art." *Diplomatic History* 9 (Fall 1985).

"The War Ends." *Life*, 20 Aug. 1945.

Wattenberg, Albert. "December 2, 1942: The Event and the People." *Bulletin of the Atomic Scientists* 38 (Dec. 1982).

Weart, Spencer R. "Learning How to Study Images as Potent Forces in Our History." *Chronicle of Higher Education*, 8 Mar. 1989.

Weinberg, Alvin M. "Social Institutions and Nuclear Energy." *Science* 177 (7 July 1972).

Weintraub, Bernard. "Civil Defense Agency Suddenly in Spotlight." *International Herald-Tribune*, 13 Apr. 1982.

Weisgall, Jonathon M. "Micronesia and the Nuclear Pacific since Hiroshima." *SAIS Review* 5 (Summer–Fall 1985).

Wells, Samuel F., Jr., "The Origins of Massive Retaliation." *Political Science Quarterly* 96 (Spring 1981).

Weltman, John J. "Trinity: The Weapons Scientists and the Nuclear Age." *SAIS Review* 5 (Summer–Fall 1985).

"West Coast Gets Ready." *Life*, 12 Mar. 1951.

Wheaton, William L. C. "Federal Action toward a National Dispersal Policy." *Bulletin of the Atomic Scientists* 7 (Sept. 1951).

[White, E. B.]. "Notes and Comment." *The New Yorker*, 18 Aug. 1945.

"Whoops! A $2 Billion Blunder." *Time*, 8 Aug. 1983.

Wicker, Tom. "Civil Defense a Dangerous Band-Aid." *Eugene* [Oregon] *Register-Guard*, 2 Aug. 1982.

———. "That Bang Resounds to This Day." *International Herald-Tribune*, 19 July 1985.

Will, George F. "An Exhibit: *Enola Gay* for the Kids." *International Herald-Tribune*, 20–21 July 1985.

Wilson, Robert R. "Niels Bohr and the Young Scientists." *Bulletin of the Atomic Scientists* 41 (Aug. 1985).

Winkler, Allan M. "American Hopes and Fears in the Nuclear Age." *Sweet Reason: A Journal of Ideas, History & Culture*, published by the Oregon Committee for the Humanities, Portland, Oregon, 5 (Fall 1986).

———. "A 40-Year History of Civil Defense." *Bulletin of the Atomic Scientists* 40 (June–July 1984).

———. "Harry S Truman and Leadership in the Atomic Age." *Power and Responsibility, Case Studies in American Leadership*. Edited by David M. Kennedy and Michael E. Parrish. San Diego: Harcourt Brace Jovanovich, 1986.

———. "The Nuclear Question." *American Choices: Social Dilemmas and Public Policy since 1960*. Edited by Robert H. Bremner, Gary W. Reichard, and Richard J. Hopkins. Columbus, Ohio: Ohio State Univ. Press, 1986.

Wohlstetter, Albert. "The Delicate Balance of Terror." *Foreign Affairs* 37 (Jan. 1959).

Wollscheidt, Michael G. "Fail Safe." *Nuclear War Films*. Edited by Jack G. Shaheen. Carbondale and Edwardsville, Ill.: Southern Illinois Univ. Press, 1978.

York, Herbert F. "The Debate over the Hydrogen Bomb." *Scientific American* 233 (Oct. 1975).

———. "Vertical Proliferation." *Bulletin of the Atomic Scientists* 38 (Dec. 1982).

Zuckerman, Edward. "Atomic Dream & Nightmare." *New York Times Book Review*, 31 Oct. 1982.

Zuckerman, Lord. "Nuclear Fantasies." *New York Review of Books*, 14 June 1984.

Index